NOTES FOR THE SHEEP CLINICIAN

by

M.J. Clarkson and W.B. Faull

Department of Veterinary Clinical Science
University of Liverpool

with contributions from

D.A.R. Davies

Department of Animal Husbandry
University of Liverpool

G.C. Skerritt

Department of Veterinary Anatomy
University of Liverpool

G.S. Walton and W.R. Ward

Department of Veterinary Clinical Science
University of Liverpool

First published 1983 by

LIVERPOOL UNIVERSITY PRESS,
P.O. BOX 147, LIVERPOOL, L69 3BX

Second edition published 1985

Third edition published 1987

British Library Cataloguing in Publication Data

Clarkson, M.J.
Notes for the Sheep Clinician. - 3rd ed.
1. Sheep - Diseases
I. Title II. Faull, W.B.
636. 3'089 SF968

ISBN 0-85323-365-9

Word-processed by the
Composing Unit, University of Liverpool

Printed by
Eaton Press Limited, Wallasey, Merseyside

This book grew out of brief duplicated notes which were given to Final Year Veterinary Students at the Liverpool University Veterinary School. Although mainly designed for undergraduate teaching purposes, where it complements a series of 20 lectures, these notes should also be of interest and value to sheep practitioners and farmers, particularly in the United Kingdom.

It begins with sections on the economics of sheep production and then works its way through the sheep year beginning with breeding and feeding, followed by sections on the disease problems of the adult ewe. These are followed by sections on perinatal disorders and on conditions affecting the grazing and fattening lamb. The book ends with notes on sheep health schemes and the systematic examination of a sick ewe. In the third edition, we have updated every section and added some new sections.

"The third time he asked, 'Do you love me?'

- 'then feed my sheep' "

John 21: 17

Acknowledgements

We are very grateful to all those who encouraged us to have these notes published, particularly to Andrew Madel and Andy Eales who originally had a look at them and made suitable comments! We are also very grateful to Joe Read for advice over the economics section. We are also grateful to many farmers who allowed us to practice on them and their sheep and to numerous veterinary surgeons, especially Agnes Winter, who shared their knowledge with us. The third edition has been expanded and modified in the light of comments on the two editions made by reviewers and by final year students. This book would never have been completed without the care and efficiency of our secretary, Jean White. We would like to thank Hilary Davies for compiling the index.

We have often given some indication of the price of drugs and vaccines, based on information in the October, 1986 Index of Veterinary Specialities. In the case of PML products, the price is as stated in IVS; with POM products we have added the usual percentage.

CONTENTS

LAMBING TIME

C.N.S. DISEASES OF LAMBS

TICKS AND TICK-BORNE DISEASES

THIN, SCOURING LAMBS

PRODUCTION

ECONOMICS OF SHEEP PRODUCTION

In order to appreciate the value of individual animals and complete flocks it is necessary to have some idea as to the main income and costs of the sheep farmer. This information will form an essential background to any veterinary involvement on the sheep farm. The farmer will tend to be more aware of obvious savings such as the value of an individual ewe which might be saved by clinical treatment but preventive health programmes are more financially rewarding to the farmer (and vet) since they influence the overall productivity of the flock.

The Meat and Livestock Commission has over 500 flocks with all systems of management on its recording scheme and publishes data on the production figures and on the finances of these flocks. They also divide the data into 'average', dealing with all the flocks of one system, and 'top-third', which shows what improvements are possible and may act as a target for other farmers. It must be recognised that the data for the 'average' flocks are probably much higher than the average of all flocks since the onerous task of keeping accurate records is bound to select for better flocks. These sheets can be obtained from MLC, Sheep Improvement Services, P.O. Box 44, Queensway House, Bletchley, MK2 2EF.

An analysis by the MLC of its records from commercial lowland flocks producing lambs for slaughter shows that stocking rate, number of lambs reared, flock replacement costs (ewe longevity), lamb sale price and feed and forage costs are main performance factors which influence the profitability of the flock.

These factors account for 93% of the superiority of the 'top-third' of their flocks, suggesting that these areas are where the greatest advice is needed. Veterinary advice can improve all these factors e.g. parasite control is crucial to increasing stocking rate and will also influence the time taken to attain marketable weight, which influences the value of the lamb sold. The value of the lamb sold can be influenced most rapidly by changing the terminal ram sire and MLC has carried out large-scale controlled trials with different rams on different commercial ewes. The ram influences carcase weight and the rate at which lamb mature to the desired fat level and reference should be made to MLC results and/or ADAS advisers before recommending a change in ram breed.

This section gives an introduction to EEC price structure, examples of physical and financial production of different types of flock and a few conclusions drawn from them.

A Layman's Guide to the Common Agricultural Policy for Sheep Meat

This policy, which commenced in 1980 in U.K., has increased the income of sheep farmers considerably and has stimulated an increase in the national ewe breeding flock. U.K. ewes at the December 1985 census number almost 16 million which produce a similar number of lambs, most of which would be eventually slaughtered for meat. An outline of the pricing structure for prime slaughter lambs is given here.

Major changes have been proposed but have not been adopted. The EEC is only about 70% self sufficient in sheep meat and still relies heavily on imports, mainly from New Zealand, to meet demand. The total expenditure on the sheep meat regime is less than 2% of the agricultural budget and, particularly in the less favoured areas, sheep production is the basic agricultural commodity. The MLC published comparative figures for the first 5 years of the sheep meat regime in 1986 which indicated that the gross margin per ewe, adjusted for inflation, has fallen between 1981 and 1985 across all major systems of sheep production, by about 3% in hill flocks and 10% in upland and lowland systems.

1. A weekly guaranteed price ('Guide-price') is agreed:

   (a) Set annually for Great Britain by the Council of Ministers.

   (b) Seasonal changes over the year are marked and thus influence farmers' marketing policy. The pattern of seasonal changes was changed in 1984 so that prices fell steeply during June and July. This meant that if farmers do not get lambs to slaughter weight by July, there was little commercial incentive to send them for slaughter until November, since the guide price remains the same throughout August and September.

   e.g. July - Sept. 1986: 206.6p/kg est. dressed carcase weight (e.d.c.w.)* lowest

March 1987: 268p/kg e.d.c.w. (highest)

* e.d.c.w. - half live weight. The 'best' live weight at which to market lambs of different breeds and crosses can be estimated as follows:-

$$\text{live weight} = \frac{\text{sum of weight of mature ram and ewe}}{4}$$

2. This does not remove the necessity to handle individual lambs but gives an indication of slaughter weight. The average U.K. market price is frequently lower than the guide price. e.g. July 1985. Guide price 200p/kg; market price 156p/kg.

Occasionally there are months when market price is just above the guide price, (usually around Easter - e.g. April 1986)

3. Difference between guide price and average market price is made up by the payment of a 'variable premium' to producers from EEC funds.

e.g. July 1985: variable premium = 44p/kg e.d.c.w.

This premium is only paid on 'certified' carcases and it is essential that all carcases should be certified for a farmer to be successful.

4. The certification system discourages fat carcases. 'Fat' carcases are 'rejected' at market by graders and the variable premium will not be added to their market price. However, the majority of carcases qualify for 'variable premium' but it is possible that the system may be altered to further favour lean carcases.

5. Exporters to EEC countries (France being most important for U.K.) have to pay a levy, which is equal to the variable premium - this is the so-called 'claw-back'. This, as well as other measures, has reduced exports to EEC countries considerably.

The 'claw-back' is not payable on exports to non-EEC countries.

6. In addition to 'variable premium', which is paid on the final product, annual ewe premiums are paid.

Two types of ewe premiums are paid:-

(a) Annual ewe premium paid entirely from EEC funds. This is paid on all ewes and depends on the annual average market price of lamb. A calculation is made in retrospect to bring the average producer's return up to a 'Reference Price' which is agreed by the Ministers at the beginning of the marketing year in April. The emphasis has been for the ewe premium to increase over the 5 years of the regime and the average ewe premium was £4.21 in 1985.

(b) Less Favoured Areas premium. The EEC recognises that certain areas (especially hill areas) are particularly difficult to farm, production costs being high and farm incomes below non-agricultural incomes. Member countries may pay a compensatory allowance to such farmers, limits being set for its amount, 25% of its cost coming from EEC funds and 75% from the member country. The annual compensatory allowances in Britain are £4.25 for upland ewes and £6.25 for hill ewes.

The variable premium scheme outlined in the foregoing does not apply to Northern Ireland in order that its scheme agrees with that obtaining in Eire where the payments are entirely on the ewe premium basis. (England, Wales and Scotland are the only EEC areas where a variable premium system is used but France is actively considering this system.)

Examples of MLC Records (latest i.e. figures from the 1985 lambing season)

Variable costs are those which can be assigned to the sheep section on a mixed farm and vary with the number of sheep kept. These are used in the calculation of gross margin. Fixed costs, which may not be readily assigned to a particular venture and are constant whatever the number of sheep kept, must be deducted to obtain net margin.

HILL (based on 30 flocks - a very small proportion so is not representative)

Physical results (per 100 ewes to tup)

Lambs born	119	(129)
Lambs reared	105	(120)
Lambs retained	23	(25)
Barren ewes	5	(4)
Dead ewes	4	(4)
Productive ewes	93	(95)

Income (per ewe to ram) £			Variable Costs £		
Sale of lambs	23.39	(31.54)	Concentrates	3.22	(2.94)
Sale of draft ewes	4.90	(6.56)	Other Purchased feed	1.24	(0.87)
Wool	1.75	(1.80)	Forage/grassland costs	2.16	(1.63)
Ewe premium + LFA subsidy	10.42	(10.84)	Vet/medicines	1.97	(2.25)
			Market/transport etc.	0.70	(1.01)
Gross returns	40.46	(50.74)	TOTAL	9.29	(8.67)
Less flock replacements (rams)	1.61	(1.91)			
OUTPUT	38.85	(48.83)	GROSS MARGIN PER EWE	29.56	(40.16)

UPLAND (105 flocks, selling lambs for slaughter and as stores)

Physical results (per 100 ewes to tup)

Lambs born	146	(160)
Lambs reared	131	(145)
% reared sold direct to slaughter	47	(51)
Lambs retained for breeding	10	(12)
Barren ewes	6	(5)
Dead ewes	5	(4)
Productive ewes	92	(94)

Income (per ewe to ram) £			Variable Costs £		
Sale of lambs/valuation	47.11	(53.31)	Ewe concentrates	5.72	(5.78)
Wool	2.40	(2.62)	Other purchased feed	2.15	(2.09)
Premium and subsidy	7.97	(7.88)	Forage/grassland costs	5.50	(5.70)
			Vet/medicine	2.64	(2.69)
			Market/transport etc.	0.94	(0.99)
Gross returns	53.86	(55.76)			
Less flock replacements	7.63	(4.16)			
OUTPUT	46.23	(51.60)	TOTAL	15.51	(15.75)
			GROSS MARGIN PER EWE	34.53	(40.75)

LOWLAND (398 flocks, selling lambs for slaughter and as stores

Physical results (per 100 ewes to tup)

Lambs born	166	(174)
Lambs reared	148	(157)
% lambs sold direct to slaughter	54	(61)
Lambs retained for breeding	2	(1)
Barren ewes	6	(5)
Dead ewes	5	(5)
Productive ewes	92	(93)

Income (per ewe to ram) £			Variable Costs £		
Sale of lambs/valuation	54.37	(58.89)	Ewe concentrates	7.26	(6.86)
Wool	2.98	(3.09)	Other purchased feed	2.15	(2.09)
Ewe premium	4.21	(4.47)	Forage/grassland costs	6.99	(6.38)
			Vet/medicines	3.28	(2.98)
			Market/transport etc.	1.28	(1.07)
Gross returns	61.56	(66.45)			
Less flock replacement	9.54	(8.13)			
				20.96	(19.39)
OUTPUT	52.02	(58.32)			
Stocking rate (grass and forage per ha.)	12.00	(14.40)	GROSS MARGIN PER EWE	31.06	(38.93)
Kg nitrogen/ha	160	(184)	Gross margin per ha.	373	(561)

The average results are quoted with the results for the 'top-third' of flocks in brackets.

Fixed costs must be deducted to obtain net margin, i.e. gross margin less labour, rent, depreciation; these vary considerably between different enterprises.

Some useful data (approximate prices)

Dead ewe (lowland)	=	£70 loss
Dead lamb "	=	£25 loss
Abortion "	=	£45
Cost of barley	=	£100/tonne
Cost of ewe/lamb concentrate	=	£150/tonne

CONTROL OF BREEDING

1. Early lambing with progestagen sponges

Discuss with the farmer when he can sell finished lambs, and whether he wants lambs born before Christmas or after. (Pedigree ram lambs should be born after 1st January.)

e.g. calculate: last lambing	- 23 December
if longest gestation 150 days, remove rams	- 26 July
if rams with ewes 4 days, remove sponges	- 22 July
if sponges in for 13 days, insert sponges	- 9 July

'Veramix' sponges from Upjohn contain 60mg medroxyprogesterone. 'Veramix Plus' are similar but supplied with pregnant mares serum gonadotrophin (PMSG) (the benefit of PMSG is probably not worth the cost when sponges are inserted in July). 'Chronogest' from Intervet contains 30mg fluorogestone acetate.

Roughly half the ewes will lamb to the induced oestrus and most have twins. Results vary widely, according to breed, feeding, time of year and weather. Test a batch of ewes and record the results in order to find out whether the procedure is worth while on a particular farm. Ewes and rams should be in good condition (Score 3) and fit. If possible, remove ewes from the ram when they have been marked (raddled) convincingly, to avoid repeated serving of favourites. Put the rams near the ewes at sponge withdrawal, and put them with the ewes 48 hours after withdrawal, for 48 hours. Two weeks after first oestrus, put rams with the ewes for one week, with different raddle crayon and then remove. The usual advice is to allow 1 ram for 10 ewes. It is possible with some individual rams to achieve good fertility with more ewes. There are big differences between individuals and between breeds in the ability of rams to breed outside the usual season. Roughly 10% of rams are of very low fertility, in any case.

Only well-managed farms can benefit from early, concentrated lambings. Marketing of lambs, and housing and labour for lambings must be planned well. MLC figures for 1977-80 show a slightly higher gross margin from early lambs (£228/hectare) than from grass lambs (£213/hectare).

In theory, ewes may be weaned early, mated in February without hormone help, and three crops of lambs produced every two years. In practice, very few people have achieved this. Dr. W.M. Tempest, of Harper Adams Agricultural College, has managed a successful system which produces lambs every 8 months from ewes mated in December, April and August. He uses Finnish Landrace X Dorset Horn ewes and Down rams. Lamb sales approaching 300%, and profitability three times higher than in annual lambing flocks are reported.

2. Early breeding with vasectomised rams

A vasectomised ram run with ewes one month earlier than the usual start of the breeding season will stimulate a large proportion of the flock to come into oestrus about 3 weeks after his introduction. This is a cheap method of inducing synchronised oestrus. The vasectomised ram may also be used to withdraw individual ewes, as they are raddled, for service by a fertile ram, or for artificial insemination.

3. Effect of shearing on start of breeding season

Clun ewes which were shorn before mid July began to cycle earlier than ewes which were not shorn until early August. (During the same period in different years a higher mean ambient temperature delayed the onset of breeding: a difference of 2.1°C (3.8°F) was associated with a difference of 20 days.)

4. Out of season breeding with progestagen sponges and PMSG

Disappointing results have been obtained earlier than July. A successful attempt was made by Liverpool University in 40 barren or aborted cross-bred ewes in North Wales. Sponges were removed and 1000 units PMSG injected on 25 May. Twenty ewes lambed, producing 30 live and 7 dead lambs and the lambs fetched a high price when sold for slaughter.

Elvidge from Oxford University has mated Mule (Blue-faced Leicester X Swaledale) ewes in June, with progestagen sponges and PMSG (Veramix Plus), with 50% of ewes becoming pregnant.

5. Out of season breeding with light control

Sheep that are housed may be given a shortening daylength, to induce cycling.

6. Batch lambing during the breeding season

Progestagen sponges (without PMSG) may be used. Prostaglandins may also be used but fertility in the induced oestrus has been poorer than after sponges. The second oestrus is well synchronised. Two doses are given 10 days apart. Estrumate (100ug cloprostenol) or Lutalyse (15mg F2α) are effective.

7. Induction of lambing

Predictable lambing dates can be achieved by injecting 8ml (16mg) betamethasone or dexamethasone 140 days of gestation or later.

In a trial in 1983 at Liverpool University, 33 ewes were selected for injection at 9 p.m. on day 142 of pregnancy (Saturday). Two lambed before treatment, 3 within 24 hours, 3 between 24 and 36 hours, and 24 between 36 and 48 hours (9 a.m. - 9 p.m.

Monday). Only one ewe lambed later, 55 hours after treatment. It is clearly possible to give better attention to ewes, and to give attendants a rest, particularly avoiding the few ewes at the end of a group, with long gestations. Side-effects appear to be negligible.

8. 'Fecundin'

Immunisation against androstenedione increases the ovulation rate, by preventing the physiological inhibition of gonadotrophin. It was marketed in 1985 in U.K. as 'Fecundin'. A 2ml dose is given 8 and 4 weeks before tupping begins. In subsequent years one dose is given 4 weeks before tupping. One ml contains 0.6mg ovandrotone albumin.

An A.D.A.S. trial of 100 ewes showed 23 extra lambs from every 100 ewes tupped. They advise using it on flocks with lambing rates between 120 and 160%. In one farm, Scottish Blackface ewes on a hill showed no response, whereas ewes on improved pasture produced 12% more lambs. To cover the costs, about 10% more lambs are needed.

Ram Examination

About 10% of rams have poor fertility. Many of these can be detected by palpation of testicles, and most by examination of fresh semen.

Examination of a ram lamb before use, or of a freshly-bought ram, is a wise investment. Ewes run with an infertile ram can lamb two months late. If several rams of the same breed work together, an infertile ram may never be detected.

Any history based on good breeding records is valuable. Illness, however brief, can lead to temporary or permanent infertility up to two months later. If the ram is examined before entry to the flock, his conformation and any inherited conditions should be checked. Teeth should make contact with the dental pad. Orf should be looked for at the commissures of the mouth. Condition score should be 3 to 3.5. Feet should be checked.

Before the breeding season (before July in the British Isles) the size of testicles and numbers of spermatozoa are lower, especially in breeds with a short season.

Palpate the testicles. They should be of similar size, very firm (turgid) and move freely in the scrotum (compare with others of the same age and similar breed). The scrotal circumference averages 38cm in mature Suffolks. Testicular hypoplasia is incurable.

The penis can be extended with care in the sitting position, to check for rare defects.

Electroejaculators can be powered by mains electricity (Semen Sampler, Centaur Veterinary Equipment, Edinburgh) or by battery (Ruakura type, Alfred Cox). The tip of the probe should be on the pubic brim, but there are marks to indicate the depth of insertion in the average ram's rectum. The stimulation of muscles looks distressing, but hearsay evidence from human volunteers is that it is not painful. Nevertheless, it seems reasonable to give up if a sample is not obtained after 3 or 4 attempts and try again later. The ram can stand or lie. The semen can be collected in a warm beaker or into a small transparent plastic bag. Between 0.5 and 2ml of dense creamy semen is normally collected from a fertile ram.

Semen examined immediately on a warm slide in a warm room should show good wave motion (as if being vigorously stirred) as seen in bull semen. If a poor sample is obtained a second sample should normally be taken. After mixing one drop of semen on a warm slide with 5 drops of nigrosin-eosin, (1.67g Eosin, 10.0g Nigrosin, 100ml water) for three minutes, a smear can be made. The total number, number live (unstained), and number free of morphological abnormalities provide further evidence of the likely fertility.

If the history and examination of the ram and semen show reduced fertility, treatment is not normally possible. A potentially valuable ram may be tested again two months later, as infertility is occasionally temporary.

A report should always remind the client that the only final evidence of fertility is the production of lambs.

Artificial Insemination

A.I. in British sheep is not yet common. In 1984 the Meat and Livestock Commission (MLC) for the first time offered a commercial insemination to 1000 ewes within 50 miles of their Pig Breeding Centre, Thorpe Willoughby, Selby. Fresh semen pooled from Suffolk rams selected for fast growth gave a non-return rate (NRR) of 71%.

The advantages for a market where little premium is paid on quality is in fast growth of lambs.

The cost, including synchronisation by intravaginal sponge, and 500 iu PMSG at withdrawal is £5. Withdrawal at 8 a.m. is followed by A.I. at 4 p.m. 2 days later (56 hours). Clearly, pooled semen is inappropriate for pedigree breeders, who can, however, obtain advice and help from MLC, including training in collection and insemination.

Semen frozen in pellets gives 50% pregnancy rate which is acceptable for export. Straws inseminated into the cervix produce only 30% pregnancies. Since the sheep's cervix is virtually impenetrable, intra-uterine insemination has been done by laparoscopy (under local analgesia). Only one-tenth of the dose of semen is needed and a higher pregnancy rate is obtained. Questions have been raised about the ethics of this procedure.

The interest in goat and sheep breeding could possibly lead to a joint A.I. service.

The potential of A.I. is high: 7000 doses of fresh semen or 2500 doses of frozen semen could be collected each year from a Suffolk ram. However, it would be necessary to have proven sires for valuable characteristics.

Diagnosis of Pregnancy and Number of Lambs by Ultrasonic Scanning

A large number of people have now set up as scanners, and an organisation called Ewescan exists to ensure quality control among affiliated users (Newton Bank, Frankscroft, Peebles). For 50p to £1 a ewe, the number of foetuses is estimated, with errors less than 2% for experienced operators.

The value of the service is that barren ewes can be sold if prices are high, or fed less until later. Ewes with single lambs can be fed less, saving feed, and avoiding over-large lambs. Ewes with two or more lambs can be fed more concentrates and watched carefully for loss of condition or pregnancy toxaemia. The Hill Farming Research Organisation (HFRO) and ADAS have estimated the value of scanning at £3 - £4.50 a ewe.

The original machines used a linear array and, for this the ewes must be shorn to 20cm cranial to the udder, across the whole ventral abdomen, and kept off roughage for 8 hours. They either lie in a cradle or sit up. One recent machine has a sector array and may be used in the standing position without shearing. For accurate diagnosis of foetal numbers the flock should be scanned between 50 and 100 days after tupping.

Other methods of diagnosing foetal numbers include X-ray (more expensive, not generally available, slightly hazardous), measurement of progesterone in blood (unreliable) and Doppler ultrasound (unreliable). The main alternative is regular condition scoring and transfer of thin ewes to groups fed more concentrates. (This avoids the need for pregnancy diagnosis, but requires constant observation by good shepherds during late pregnancy.)

FEEDING MANAGEMENT

Grass, either fresh or conserved is the main feedstuff for sheep in the United Kingdom. However, because of seasonal effects and varying requirements during the production cycle, there are periods during the year when grass needs to be supplemented with concentrates.

Feeding the Ram

Rams must be in fit condition 6 - 8 weeks before mating with a condition score of 3 to 4. In many circumstances, this will be achieved by providing reasonable grazing. If grass is insufficient then concentrates may need to be fed in the two month period prior to and during the mating season. Up to 400g/day of a 16% crude protein concentrate is advised but overfeeding should be avoided. Frequently, rams are neglected and are in poor condition not only because of inadequate feeding but also because of parasites, feet and other health problems.

Feeding the Ewe

There are three important periods when grass supplies may need to be supplemented.

(1) pre-and immediate post-mating
(2) late pregnancy
(3) early lactation

1. Pre- and immediate post-mating

Ewes in below optimum condition at this stage will have lower ovulation and conception rates and increased likelihood of barrenness. The optimum condition score range for hill and mountain ewes is 2.5 to 3.5 and for lowland ewes 3.0 to 4.0. Ideally, ewes should gain weight in the pre-mating period, be in good condition at mating and remain so for a month post-mating.

Adequate amounts of grass of reasonable quality must therefore be provided for two months in the late summer and autumn which necessitates the farmer reserving an area of pasture for this purpose. In some situations, more especially in later lambing flocks, grass may have to be supplemented with concentrates; the actual amounts fed will, however, vary considerably and will be dependent on grass availability and ewe condition. Poor condition ewes need careful investigation - the presence of significant numbers is usually associated with underfeeding and overstocking. However, poor condition may also be attributable to health problems such as faulty teeth or fascioliasis (see later).

2. Late pregnancy - 8/6 weeks pre partum

This coincides, in most cases, with winter months when there is little grass available of inferior quality. Conserved forage is therefore the main source of nutrients with concentrates provided when necessary.

The amount of concentrates to feed depends on:-

(i) quantity of forage available

(ii) quality of forage available (any grass available will be equivalent to medium or poor quality hay).

The quality of conserved grass will be determined by the dry matter percentage (DM%) in the case of silage and in both hay and silage by the digestibility of the organic matter in the dry matter ('D' value), by the metabolisable energy concentration (ME value in Megajoules per kilogram) and by the crude protein content (CP%). These may be obtained directly or indirectly by chemical analysis of the foodstuff (see later). A good guide on the farm can be obtained by assessment of the stage of maturity when the grass was conserved, and, for silage, by estimating the dry matter percentage (squeeze) and type of fermentation (smell). Grass cut early will be leafy, have few seed heads and will produce high quality forage; on the other hand, poor quality material will result if cutting is delayed far beyond the emergence of the inflorescence and will consist mainly of stalky lignified stems and seed heads.

(iii) live weight of the ewe

(iv) condition of the ewe

The condition scores range from 1 thin to 5 fat. Scoring at 8 - 6 weeks prelambing which may coincide with housing and 4 - 2 weeks pre partum, when vaccinating for clostridial disease, will provide a good guide to the adequacy of feeding. Lowland ewes should have a condition score of 3.5, upland or mountain ewes 2.5 or more at both times. If ewes have been sheared then changes in condition score can be easily observed.

(v) foetal load

This can be estimated on the basis of previous performances of the flock and the estimate may have to be modified in individual cases in the last weeks of pregnancy if the condition score changes rapidly. Ultra sound scanning by trained operators in the period between 50 and 100 days of pregnancy gives a very accurate prediction of the number of lambs being carried.

(vi) stage of pregnancy

The use of ram raddles at mating time gives extra precision to this.

(vii) appetite level

Dry matter intake is approximately 1.5% of live weight in early and mid pregnancy but increases in the latter stages to 2.0 - 2.5% until the last days, when there is a marked decline. There is, however, considerable variability; the digestibility of the ration will have a major influence with poor quality, high fibre diets causing physical restriction and reduced appetite.

Steps in the construction of a suitable ration for a 70kg ewe of condition score 3.5 carrying twins from 8 weeks pre partum

(i) Obtain an estimate of the quantity and quality of foods available

Compositional quality of some common foodstuffs - figures are on oven-dry basis for ME, protein and mineral levels.

		DM%	ME MJ/Kg	Crude Prot. %	Degradability of Prot. %	Ca%	P%	Mg%
Hay,	medium quality	86	8.7	9.0	80	0.3	0.2	0.08
	good quality	86	9.5	11.0	80	0.3	0.2	0.08
Silage		25	10.5	14.5	80	0.3	0.3	0.16
Concs.	oats	87	12.5	10.0	80	0.08	0.35	0.1
	barley	87	13.0	9.5	80	0.06	0.4	0.1
	Soya bean meal	90	12.2	45.0	60	0.3	0.75	0.3
	fish meal	90	11.0	63.0	40	6.0	3.8	0.2

Typical mineral vitamin mix composition: Ca = 15% P = 8% Mg = 12% NaCl = 15% Mn = 4000 ppm Fe = 3000 ppm Zn = 1000 ppm I_2 = 40 ppm Co = 20 ppm Se 10 ppm Vit A = 400 iu/g Vit D_3 = 40 iu/g Vit E = 0.8 iu/g

Note that-:

(a) The most variable feedstuffs are the forages and there could be some value in having particular hay and silage samples analysed chemically. The ME concentration can be calculated from the fibre level in the foodstuff by the following formula:-

Hay ME (MJ/Kg DM) = 16.5 - 0.21 x Modified Acid Detergent (MAD) fibre

Silage ME (MJ/Kg DM) = 14.6 - 0.13 x MAD fibre

The ME value can also be predicted directly from 'D' values by the equation ME = 0.15 D

(b) The crude protein content can also be obtained by analysis. However, the figure obtained may be inflated because the analysis and calculation assumes all the nitrogen measured has a protein source.

(c) Compound foods may also vary. The ME value of such foods can be predicted from the declared analysis by the following equation:

* ME = 12.0 + 0.08CP + 0123 EE - 0.18 CF - 0.12 TA
CP = crude protein, EE = ether extract, CF = crude fibre, TA = total ash

(All analytical values are expressed as % ages in the dry matter)

(d) The degradability of the protein is a measure of how much is likely to be degraded in the rumen - RDP and how much will escape microbial action and pass directly into the abomasum - UDP.

(e) In some cases, the source of protein and minerals and vitamins will be a 'supermix' product. This is often more convenient for the farmer but the information of the mineral/vitamin levels for specific ingredients is not often readily available. A straight mineral/vitamin mix will usually have such a detailed analysis. The analysis varies from product to product so it should be possible to choose one which will meet specific needs. It should be remembered that consumption depends not only on the inclusion rate but also on the amounts being fed.

[* Source MAFF Reference Book 433, Energy allowances and feeding systems for ruminants]

(ii) Estimate ewe requirements and appetite per day

Weeks pre partum			8	6	4	2	0
Appetite	Dry matter (oven-dry)	(Kg)	1.3	1.35	1.5	1.65	1.3
Requirements	ME (MJ)		11	13	15	17.5	20.5
	C. Protein RDP	(g)	110	140	170	210	250
	Ca	(g)	4	6	6.5	7	8
	P	(g)	2	3	4	4	5
	Mg	(g)	1.5	1.5	1.7	1.7	2.0
	NaCl	(g)	2.0	2.0	2.5	2.5	3.0

Note that:-

The above figures have to be adjusted when ewes of different weights and condition scores are being considered and whether ewes have a higher foetal load. The approximate adjustments suggested are a 10% change for a difference of 10kg live weight or 1 unit of condition score, and a 10% increase for ewes carrying triplets.

(iii) Calculate amounts to feed g DM/day

e.g. with the following available foods - moderate quality hay, barley, soya bean meal and sheep mineral mix

Weeks pre partum		8	6	4	2	0
Hay		1500	1250	1240	1100	250
Concentrates -	Barley		235	380	600	760
	Soya bean		50	100	175	350
	Vit. Min. mix		15	20	25	40
	TOTAL		300	500	800	1150

Note that:-

(a) The amounts shown above are for on the farm air dry materials. It is necessary, therefore, to include a factor for DM% in the calculations e.g. @ 8 wks an intake of 1.5kg of air dry hay is equivalent to $1.5 \times \frac{86}{100} = 1.3$kg oven dry material.

(b) As much forage as possible is included to minimise costs. Hay levels are approximate and reflect appetite; it is important to establish that the animals are actually eating these amounts especially with poor quality material with which wastage can be considerable.

(c) In the last days of pregnancy requirements are not being met so the ewe will lose condition. Feeding higher quality hay would improve the situation but increasing the concentrate intake could cause metabolic upsets (see cereal over-eating notes). Feeding whole barley reduces the risk of this occurring as does the feeding of a complete diet.

(d) Pregnancy toxaemia may occur if only good quality forage is provided. This is because, although total energy is sufficient, there is a lack of glucose energy which is normally synthesised from propionic acid. Including a supply of readily digestible carbohydrates overcomes the problem.

(e) If the farmer wishes to adjust feeding every week it is sufficient to recommend that concentrate feeding is at the mid-point of the foregoing figures (iii) for intermediate weeks.

(f) The consequences of under nutrition are smaller lambs, higher mortality, poorer milk yield and increased body condition loss and incidence of pregnancy toxaemia. The farmer, however, may accept lower performance if this means substantial cost saving; in particular the protein level may be deliberately kept below the above. Economies made at this stage can have long-lasting effects and are usually ill-advised.

'Flat rate' feeding

Flat rate feeding concentrates is an extreme modification to the steadily increasing rationing system outlined above. This system involves calculating the total concentrate need for a particular period and then dividing into equal daily aliquots for the whole period which makes daily feeding easier. As a result, ewes increase in weight and condition in the early part of each period and lose considerable amounts in the later stages. The full effects of this can be reduced to some extent by changing the quality of roughage fed which should be available ad libitum.

3. Early Lactation

(i) The same principles of ration formulation apply. Spring grass is, however, often available to meet part or all of the ewe's needs.

Requirements and appetite level per day for 70kg ewe of CS 3.5 suckling twins each growing @ 275g/day:

Weeks post partum			1	2	4	6	8
Approximate milk yield (litres)			2.5	2.75	3.0	2.75	2.0
Appetite DM Kg concs. + hay			2.0	2.35	2.6	2.6	2.2
or grass			1.6	2.2	2.8	2.8	2.6
Requirements	ME MJ		26	30	32	30	24
	RDP)		240)	260)	270)	250)	190)
	UDP)		100)	110)	130)	90)	30)
	) CP	(g)	)340	)370	)400	)340	)220
	Ca	(g)	12	14	14	14	12
	Mg	(g)	3.5	4.0	4.5	4.0	3.0
	P	(g)	9	10	10	10	9
	NaCl	(g)	5.0	6.0	6.0	6.0	5.0

Note that:-

(a) The milk yield of ewes may be estimated by assuming that the dry matter percentage of ewes milk is 20% and the food conversion ratio of dry matter of milk to live weight gain is 1:1. In the situation specified above the estimated milk yield is therefore 2 x 275 = 550 x 5 = 2.75 litres/day.

(b) Appetite increases up to the 5th or 6th week of lactation although milk yield peaks at the 3rd or 4th week.

(c) Protein requirements are based not only on milk yield but also on likely body weight changes of the ewe.

(iii) Amounts to feed g/day (air dry materials)

Weeks post partum		1	2	4	6	8
Hay (good quality)		700	800	900	1360	1250
Concentrate	Barley	1200	1500	1750	1500	1100
	Soya bean meal	270	270	270	150	100
	Fish meal	100	100	100	50	
	Vit. Min mix	30	30	30	40	50
	TOTAL	1600	1900	2100	1800	1250
Body weight change		-	-	-	=	+

Note that:-

(a) Some loss in body condition will occur in early lactation. Provided ewes are in condition score 3 or better, (lowland ewes) or 2.5 or better (hill) a loss of 1 condition score is unlikely to affect milk yield.

(b) When little or no grass is available it is necessary to feed good quality hay in the first month of lactation.

(c) Fish meal is included because of its low degradability and beneficial effect on milk yield in the first weeks.

(d) If grass is available the concentrate can be reduced and the protein content decreased; in this case it is necessary to increase the mineral content.

(e) If concentrates are not needed on grass then some provision should be made for ewes to receive magnesium; up to 6g calcined magnesite is recommended.

(f) Feeding hay and concentrates alone is unusual after 4 weeks.

Feeding the Lamb

The ewe should supply sufficient milk for twin lambs to survive and grow well. Milk is the only source of nutrients to 3 weeks of age and the main supplier to 8 weeks. From week 4 onwards the lamb will eat measurable quantities of grass and/or creep feed concentrate if milk supply is not in abundant supply e.g. in the case of triplet lambs reared on the ewe.

Most pre-lambing mortality occurs in the first few days of life and the major cause is starvation. Insufficient milk may be due to the ewe's not allowing the lamb to suck or to a lack of supply due to poor nutrition during pregnancy or to udder problems. Starvation after the first few days is usually the result of poor feeding of the ewe in lactation or to sore teats or mastitis.

Providing the lamb with a supply of colostrum is essential. Ideally, the lamb will receive this from its dam within a few hours of birth and will continue to suck its dam and receive further amounts in the following 48 hours. Lambs will consume more than 500ml in the first day but 100 - 200ml is considered enough to give disease protection provided the colostrum is from the first two withdrawals of milk. If lambs are weak or slow it may be given via a stomach tube.

Lack of supply may be overcome by collecting surplus colostrum from ewes and deep freezing until required. Alternatively, cow colostrum can be used; however, this will have no clostridial antibodies unless the cow has been previously vaccinated. Vaccination should be done at 3 months, 1 month and 2 weeks pre-calving. It is essential that the milk from the first two withdrawals is collected. It can then be deep frozen in aliquots of 200ml. The value of colostrum to the lamb declines after 1 day of age but there will be some absorption of immunoglobulin after this time especially if the lamb has been partially starved in the first 24 hours of life.

After 2 days, lambs which are surplus to the ewe's rearing capacity can be reared artificially. The following is an outline of an efficient and economic way of accomplishing this:-

Day 2:

Introduce milk replacer. It is best to have a product with 30% fat in the powder and to mix it in the ratio by weight of 1:4 with either warm or cold water. Initially, lambs take more readily to warm milk but it is possible to use either a warm milk restricted feed 3 feeds per day system or to feed milk ad libitum. Intakes in the latter system will reach 2.5 to 3.0 litres, whilst on restricted feed 1.5 litres per day is considered the maximum to feed. Lambs on restricted intake can be fed with a bucket with teats at the base, but when feeding is ad libitum the teats need to be above milk level with a polythene tube into the milk. A food conversion of 1:1 i.e. 1kg milk powder producing 1kg live weight gain, can be expected.

Day 14:

Introduce solid food and feed ad libitum. A pelleted complete diet is preferable which includes 15% coarsely milled straw approximately 70% barley and good sources of UDP such as soya bean and fish meals to give a 16% crude protein. The mineral content should be sufficient to provide for healthy growth but should include little or none of a mineral source of phosphorus or magnesium to avoid urolithiasis problems. If coccidiosis is a likely problem a coccidiostat such as Monensin sodium can be included.

Day 20:

Provide clean water.

Day 25:

Reduce number of feeds to 2 on restricted feeding system.

Day 30:

Wean if on restricted feeding, restrict feeding if on ad libitum milk system.
It is essential that solid food is kept clean by preventing lambs walking and defaecating in feed troughs.

Day 42:

Wean ad libitium fed group.

Day 80:

Ewe lambs for breeding can be turned out to grass, preferably 'clean' and fed restricted amounts of concentrates; at first 400g/day should be allowed but this should then be decreased to 200g/day after a few weeks.

Day 100 - 120:

Lambs will be ready for slaughter at 35-40kg having grown at 300 - 400g/day from weaning and having consumed 85 - 100kg complete diet.

Lambs may also be removed from their dams at any time after they have reached 5 weeks of age and fed the complete diet. Performance figures should be similar to those stated above.

THIN SHEEP

THIN EWES AND CONDITION SCORING

Thin ewes are interesting for the following reasons:

1. They are common which should encourage us to do something about them.

2. The causes are not infinite, in fact, for most cases they are surprisingly few, for example:-

   lack of food

   inadequate teeth

   chronic lameness, chronic fascioliasis, chronic pneumonias

   This should encourage us to diagnose them accurately.

3. If there are many thin ewes, look first at the food and the number of foetuses; if there are a few, look at the 'fangs' (teeth) and the faeces; if there are one or two, look at the limbs and the lungs.

4. Thin ewes have suffered an insult for some time, which may imply irreversibility, i.e. they may not get better.

5. Thin ewes aren't "worth" much, but deserve a lot.

6. Thin ewes are generally unprofitable, for the following reasons:-

   (a) if at tupping, they produce fewer lambs

   (b) if in late pregnancy, they produce smaller lambs and risk pregnancy toxaemia

   (c) if at lambing, they produce less colostrum and show less interest in their lambs, which then in turn are susceptible to disease.

(d) if in early lactation, they produce less milk leading to poor growth rate in their lambs.

Condition Scoring (CS)

One of the reasons for thin ewes being common is because the fleece masks their condition and it is only when the ewe is shorn or handled that the thinness is disclosed. Condition scoring was introduced to overcome this problem and is now an essential tool in good flock management as well as in the clinical examination of individual sick sheep.

The crucial element in the technique is to decide by feeling the mid-lumbar region, whether the ewe is "too thin" or "too thick" i.e. is it too bony or too fat, or just about right, so that action follows, for example, by separating and supplying extra, or sometimes, less food.

A number score (1-5) is given and is rather like marking exam papers. Score 3 is mid-way (50%) and is the pass mark but with nothing to spare. Score 1 fails badly because it is much too thin and score 5 is excessive! There are always disputes between markers, but they shouldn't differ by more than half a score. It needs practice across the full range of scores and standardising with other scorers from time to time, and each time it usually requires doing a few 'to get your hand in', but it is a technique that is learnt quickly and needs no kit, only one sensitive hand! There is a useful Trainee Guide published by the Agricultural Training Board, Beckenham BR3 4PB, and another good leaflet produced by the Meat and Livestock Commission, Milton Keynes, MK2 2EF.

Two vital times to condition score a flock are 6 to 8 weeks pre-tupping and 6 to 8 weeks pre-lambing, because then there is time to alter events by, for instance, re-allocating resources such as food, housing and shepherding. Repeat scoring in about 4 weeks is advisable because at least it will suggest what the future holds, although it may be too late by then to alter some of the events. Ewes with condition score of less than 3 should be separated and clinically checked. Most will require more and perhaps better food, whilst some will require treatment or even culling.

TEETH AND AGING

It is very important to be able to examine sheep teeth properly not only for aging but also because faulty teeth are one of the most common causes for thin adult sheep. Tooth loss is also a major reason for culling sheep.

Aging - a rough and ready guide

Age (years)	
Up to One (lamb, hog, wether)	4 pairs (8) Temporary
One to two (shearling/yearling)	1 pair Permanent (2 tooth) 3 pairs Temporary
Two to three	2 pairs Permanent (4 tooth) 2 pairs Temporary
Three to four	3 pairs Permanent (6 tooth) 1 pair Temporary
Rising four to five	4 pairs Permanent (8 tooth = Full mouth)
Over five	Less than 4 pairs Permanent (Broken Mouth)

Note:-

That there is a lot of variation within and between flocks, in particular:

(1) 8 worn permanent incisors can look like 8 temporaries (some 4 year old sheep can look like less than 1 year old, without checking general features, size, teats etc.).

(2) Age of the sheep in relation to the time of year, e.g. in the autumn most sheep will be X years plus 6 months, whilst in the spring they will be just X years.

(3) Periodontal disease causes premature aging and sheep then look older than they really are.

Teeth Problems

If a few adult sheep in a flock are thin, look carefully at their teeth (if lots of adult sheep in a flock are thin, look carefully at their food!). Usually the cheek teeth cause more problems than the incisors, as they are required both for mastication and re-mastication. Make sure the ewe is facing the light (and have a good torch), and also have a good gag, e.g. block of bevelled hard wood 12 x 2 x 2cm (Shaw gag) or a metal sheep gag (Alfred Cox = £10.32).

Clinical Signs

(1) Incisors - Loose, broken mouth, uneven wear, mal-eruption, mal-position in relation to dental pad (over or undershot). Any/all of these may cause troubles directly (e.g. thin sheep) but it is more likely that they are acting as markers of more serious disturbances in the cheek teeth. They may even appear healthy, but the cheek teeth are not, and thus they can be misleading.

(2) Cheek teeth (require good gag and good light, and mind your fingers!)

Before you look inside:-

(i) Note C.S., dribbling, staining of lips and mouth, mouthing (when trying to eat), quidding, swelling of cheeks with wads of food.

(ii) Feel along the outside of the cheek for evidence of pain (flinching on pressure), and for irregularity of teeth (Shear mouth and wave mouth).

(iii) Feel with finger and thumb along the two rami for bony swellings.

(iv) Note abdominal distension for ruminal 'impaction' and/or pregnancy (likely to develop pregnancy toxaemia).

(v) Note maxillary and mandibular sinuses.

Now look inside, and with the gag, note any gaps, irregularities, loose teeth, food impaction, spikes (outer edge of upper and inner edge of lower), and tongue and cheek ulcers. Look particularly in mid-third region; it is here that loose teeth and food impaction is most common.

Causes

There is a serious progressive gingivitis following tooth eruption leading to subgingival plaque formation, periodontitis and even abscessation, together with rumen 'impaction' (rumen contents poorly masticated plus excess liquid, giving the appearance of 'water belly').

To some extent teeth eruption always causes a local 'itis' but the reason for such serious progression in some sheep and flocks remains uncertain.

Treatment

(1) Assess whether it is worth doing anything. Cull?

(2) Remove obvious loose teeth under light anaesthesia (e.g. Sagatal, May and Baker approx. 1ml/5kg.). Ewes can thrive, given a chance, without any incisors ('gummers'), providing their cheek teeth are health. Molar rasping may have short-term advantage e.g. fatteners or late pregnancy.

(3) Feed concentrates and high quality roughage.

Control

(1) Check feeding and breeding. Faults in Ca/P, Vitamin D, Copper and Fluorine levels in the feed should be considered, as well as generally inadequate feeding which may lead to excessive soil eating (which

compounds the problems of mineral deficiencies and gum irritation). Hereditary factors, including overshot/undershot mouths, may suggest changing the ram(s). Feeding blocks and roots may not promote incisor loss, but certainly require healthy teeth.

(2) 'Bite correction', sometimes called 'teeth grinding': This has been introduced to conserve overgrown incisors for an extra 1 or 2 years. Full-mouthed ewes (3 to 4 years) with incisors overshooting the dental pad, are selected for correction. The tips of the incisors are ground and re-shaped jsut enough to allow them to meet snugly on the dental pag, using a battery-operated, high-speed grinder and special gag (Alfred Cox). The technique is quick (30-50 ewes/hour), appears painless and priced at approximately £1/ewe. It looks an interesting technique but its efficacy and welfare require long-term evaluation. This technique has been declared illegal (1986) until this evaluation has been done.

(3) 'Ewe splints': The ewe splint is a stainless steel brace in acrylic resin, fixed by cement on to the incisors of ewes that are soundly full-mouthed. Ewes need careful selection and skilled application, and costs £5 - £10/ewe. Their value is questionable. (Ewesplint Ltd., Edinburgh).

FASCIOLIASIS

Fascioliasis is of immense economic importance in sheep throughout the wetter western parts of Britain. No effective immunity develops and thus disease can occur in any age of sheep. A high proportion of the flock are usually effected.

Acute disease, associated with migrating flukes is most commonly seen from October to February after a wet summer, and chronic fascioliasis, in which adults are found in the bile ducts, is seen from December to April. A sub-acute form of the disease has been described as occurring in December and January when a mixture of adult and immature flukes will be present. The best way to decide if a farm has liver fluke is to take faeces from 10 - 20 untreated thin ewes in December to May and look for eggs, though eggs are likely to be present all the year in untreated ewes.

Acute fascioliasis

Sudden death may occur in epidemic years and the mortality can be very high. Affected animals may be seen with extreme weakness, severe anaemia (PCV 10%), abdominal pain and ascites. The liver is enlarged with haemorrhagic tracts. Eggs are not found in the faeces and there are usually over 1000 immature (0.5cm long) flukes in the substance of the liver. Acute disease is seen about 6 weeks after the ingestion of a large number of metacercariae.

The flock should be treated (see later) and moved to a clean pasture, wherever possible. The metacercariae will remain viable for several months, especially over the winter so do not return sheep to the pasture containing an extensive habitat before the end of May.

Chronic fascioliasis

Affected animals show a progressive loss in condition proceeding to emaciation and signs of anaemia - pallor of mucous membranes, ascites, 'bottle-jaw' and 'flukey-eye' may be present. PCV is usually about 15%. It has been shown that even in light infections, wool yield is reduced, as is the milk production of ewes, thus resulting in reduced weight gains in the lambs. The liver is small and cirrhotic, the bile ducts enlarged and thickened. There are usually about 250 adult flukes present and eggs can be found in the faeces. The prepatent period is about 10 weeks.

Affected animals can be treated with a wide range of drugs (see later).

Prevention

This is based upon a knowledge of the environmental conditions which are needed to allow the development of the fluke from egg to infective metacercaria, on the ground and in the snail, Lymnea truncatula. This mud snail measures about 7mm and the operculum is on the right side when held upright.

The most important cycle of infection involves the 'Summer' infection of snails which results in the main pasture infection with metacercariae occurring in September and October. The 'Winter' infection of snails occur when snails become infected in October of one year and the life cycle completed in May and June of the next year. This can give rise to acute fascioliasis in July and August rather than in the autumn. It is a rare occurrence in Britain because May is usually a dry month and the over wintered snails commence to die off at this time.

Short-term control

The MAFF forecast should be used to avoid acute fascioliasis in epidemic years by preventing sheep ingesting large numbers of cercariae from September onwards by:-

(1) Removal of sheep to a pasture with no snail habitats.

(2) Fencing of small habitats.

(3) Molluscicide treatment - n-trityl-morpholine - 'Trimor', Phosyn Chemicals Limited, Pocklington, York (cost £10/acre but is quite economical for small habitats with a knapsack sprayer).

Long-term control

This is based on:-

(1) Improved drainage to eliminate snail habitats. ADAS drainage officers will survey and produce plans and grants are available for drainage.

(2) Prophylactic use of anthelmintics to reduce pasture contamination with eggs so that the proportion of infected snails is reduced.

(3) Use of molluscicides to limit snail populations, in May/June and/or August (see below).

On many farms where habitats are few and small, eradication is possible by a logical combination of these principles.

A survey of snail habitats is essential before any of these control schemes can be developed; this may be done by ADAS drainage officers or by the local veterinary surgeon with a 6" map of the fields and an experienced eye for suitable snail habitats. Improvement of hill grazings with basic slag and lime increases the pH and allows previously unsuitable land to become very suitable for snail multiplication. Severe outbreaks of fascioliasis have been seen on these improved pastures where previously the acid pH of the peat prevented extensive habitats.

Use of anthelmintics (see Table later)

Modern anthelmintics act against immature stages as well as against adults. Some drugs, such as diamphenethide and triclabendazole are effective against very early stages (1 or 2 weeks old), others against somewhat older, immature flukes (rafoxanide, brotianide and nitroxynil - 4 to 6 weeks old) whilst some drugs are only effective against adults (oxyclozanide, albendazole). Therefore, diamphenethide and triclabendazole must be used in the treatment of acute fascioliasis. Drugs are also a very important part of any control programme, the aim of which is to prevent fluke eggs being passed by the sheep (or cattle) so that the infection rate of snails is greatly reduced. Field trials with rafoxanide have shown that if all adult sheep are treated in mid-April, followed by a second treatment six weeks later, the summer infection is reduced to a very low level. Further treatments in October and January reduced fluke burdens over a 3 year period to about 10% of pre-trial numbers, and resulted in considerable increases in wool yield and lamb productivity. (The extra revenue from fleece sales was sufficient to pay for the anthelmintic treatments). This work was done in an area where fluke was widespread and habitats extensive and in areas of lighter infection, it would probably be sufficient to treat in April, October and January. The advent of the new drug triclabendazole should allow similar results to be obtained with dosing at less frequent intervals (8 - 10 weeks).

JOHNE'S DISEASE

This is an uncommon disease in sheep. It needs to be on one's list of causes of thinness in old ewes, although food, teeth, fascioliasis, chronic pneumonias and lameness are much more likely causes. However, it can become established in a closed flock, and a 10% incidence has been recorded in some flocks with an even higher prevalence in some abattoir surveys.

Diagnosis

The disease is similar to that in cattle but with the following variations:-

(1) Scouring is not so obvious, and may be intermittent with only soft faeces.

(2) As with most debilitating diseases of sheep, the fleece pulls out easily (wool break).

(3) As well as the bovine strain of Mycobacterium johnei there is also the yellow pigmented strain which can be seen at PME staining the intestine.

(4) The organisms are often not easily found in the faeces, and a necropsy and gut histology is required to establish the cause.

(5) Individual complement-fixation tests are unreliable and really only for use as a flock test.

The typical story is of a few old, dirty-tailed and broken-woolled ewes which are unaccountably thin (feed and teeth O.K.), deteriorating soon after lambing.

Control

(1) Cull clinical cases and fatten their lambs for slaughter.

(2) Vaccinate any lambs kept for breeding with 0.75ml of Ministry cattle vaccine.

(3) Ensure good control of other debilitating conditions e.g. food, worms, fluke and cobalt which appear to predispose to Johne's disease.

C.N.S. DISEASES OF ADULTS

SOME QUESTIONS OF HISTORY

How many are affected today?

Describe the signs

Lambing date (if any)

What food are they having e.g. silage, concentrates, and how much?

Have they been moved some distance in the last day or so?

What has the weather been like?

Is there shelter?

Is the grass lush?

How many cases have there been like this?

Have they been given copper? What? When?

Have the ewes been vaccinated against clostridia, louping ill - What? When?

Are there ticks on the farm now?

Have the signs improved/worsened over the last few hours/days?

What have you treated them with?

Are there bought-in ewes? If so, when?

Have any aborted - if so, how many and when?

What is their Condition Score?

Are they likely to be carrying twins/triplets?

Ewes: (producing affected lambs)

Were they purchased this year?

Did they mix with the main flock before tupping?

Have they lambed before?

Rams: (sires of affected lambs)

What rams fathered the affected lambs?

Have there been any problems with them before?

Are they related to the dams?

Were any of them bought-in last tupping time?

A CLINICAL APPROACH TO THE 3 'METABOLIC DISEASES' OF EWES

(PREGNANCY TOXAEMIA, HYPOCALCAEMIA, HYPOMAGNESAEMIA)

General

These are a notoriously difficult group of conditions for the G.P. to cope with, mainly because they are difficult to differentiate clinically and they may co-exist, the farmer has often 'got there' first and confused the picture with various treatments as well as causing delay, a number of cases may occur at the same time, the results of treatment are often disappointing and prevention is not simple on the farm. However, if one tries to keep one's feet on the ground, some sort of logic can prevail which should lead to more satisfaction for the three parties.

Clinical Diagnosis (from USUAL findings)

Pregnancy Toxaemia (PT, Twin-lamb disease)

(1) Last month of pregnancy and before spring grass.
(2) Thin (C.S.2) and often old. (Less commonly overfat suddenly starved e.g. snow).
(3) Insufficient concentrate (starving).
(4) Large abdomen (2 or more lambs).
(5) Separated, easy to catch, blind and standing motionless, fine tremors leading to convulsions (on handling). Refuse feed.
(6) Urine (by smothering) - rapid and strong positive Ketotest suggests PT; negative makes it unlikely.
(7) If not treated, progressively worsen over days and dies.

Hypocalcaemia

(1) Usually but not invariably in late pregnancy rather than early lactation, in contrast to cattle.
(2) Often been moved and not immediately fed cf. transit tetany; and therefore a number may be affected simultaneously.
(3) Often typical signs of 'milk-fever' - ataxic, leading to recumbency, depression and atony (bloat, no faeces etc.) but some are excited, hence confusion with hypomagnesaemia.

(4) If not treated, progressively worsen over hours.

Hypomagnesaemia

(1) Nearly always post-partum (lactating) and either on lush grass (plenty of food) or bare pastures (insufficient food).
(2) Excitable, leading to convulsions and rapid death.
(3) Sometimes co-existing with hypocalcaemia (hence the confusion).
(4) If not treated - progressively worsen over minutes!

Pre-treatment blood (where possible) can help retrospectively, but in practice, the diagnosis is more usually 'confirmed' by response (if any) to therapy.

Treatment

Needs to be prompt for all 3 conditions otherwise the prognosis is hopeless; hence farmer therapy.

Pregnancy Toxaemia

(1) First day - i/v, 50-100ml glucose 40 orally, 50ml propylene glycol (e.g. Ketol, Intervet) or 150ml glucose electrolyte solution (e.g. Liquid Lectade Beechams) and repeat in a few hours.

(2) Subsequently - if better, continue with glucose/glycerol until feeding well.
- if not better, either slaughter or if less than 1 week off term remove the lambs by corticosteroids (16mg) or caesar (if lambs alive).

Hypocalcaemia

20ml Ca 20 i/v (where possible) and 50-100ml Ca 20 s/c.

Hypomagnesaemia

20ml CaMgP i/v (where possible) and 50ml $MgSO_4$ 25 s/c. N.B. NOT MgSO4 25 i/v.

Control

Pregnancy Toxaemia

(1) Supply more concentrates to all sheep below C.S.3. (split and remove shy feeders) and reduce the hay. If time, do this gradually, but if emergency, consider materials like 'Glucomet' (Farm Feed Formulation Ltd.).

(2) Green grass where possible.

(3) Shelter.

(4) In future, regularly C.S. in last 2 months of pregnancy, splitting off thin sheep for special attention.

Hypocalcaemia

(1) Watch flock after movement and have Ca injection handy.

(2) Ensure hay and concentrate available (on arrival).

(3) Increase the calcium and the Vit.D in the ration during the last 6 weeks of pregnancy.

Hypomagnesaemia

(1) Watch flock after movement onto lush or bare fields and have $MgSO_4$ injection handy.

(2) If on lush pasture, move to poorer, until Mg supplement added.

(3) Feed extra magnesium (up to 14g MgO per day) via:-

- (i) Magnesium enriched cake
- (ii) Magnesium enriched feed blocks or molasses licks (less reliable than (i))
- (iii) Magnesium bullets (Rumbul Sheep/Agrimin); expensive (54p))

(4) Extra food for those underfed.

(5) Shelter.

Summary

(a 'rough and ready' approach for first visit)

(1) Pretreatment bloods (green and grey tubes) for Ca/Mg/glucose/ketones.

(2) i/v - 20ml Ca, Mg and dextrose (e.g. Ca 20 MPD solution).
(3) s/c - 50 - 100ml Ca, Mg (Ca 20 MP solution).
(4) Oral - 50 ml propylene glycol bid (e.g. Ketol) or glycerol.

Note: If the ewe is not pregnant, (2) can be Ca & Mg only and there is no need for (4).
If you are reasonably certain that it is hypomagnesaemia then (3) can be Mg only and (4) omitted.
If you are reasonably certain that it is hypocalcaemia alone, then (2) and (3) can be Ca 20 and (4) omitted.

(5) If not much improved in a few hours (perhaps a day or two for PT), prognosis grim, recheck the diagnosis.
(6) Introduce control measures.

LISTERIOSIS

This is a sporadic disease of sheep (and less commonly, cattle) seen in one or other of the following forms:-

1. Encephalitis - by far the most common and significant form
2. Abortion
3. Diarrhoea and septicaemia
4. A few reports of kerato-conjunctivitis and also mastitis
5. Septicaemia and death in young lambs

There has been an increase in reported incidents of listeriosis in the U.K. from 44 in 1975 to 272 in 1985, and is now one of the most common causes of sporadic CNS disease in individual adult sheep, particularly in the late winter months when silage is being fed and teeth are changing. It is assumed that silage feeding has much to do with this increased incidence, for it is well established that poor quality silage, particularly if it is contaminated with soil, and stored where air can get at it, can contain large doses of Listeria monocytogenes. It is thought that the organism gains entry via any mouth lesions e.g. changing teeth, and travels up to the V and/or VII cranial nerve to involve the meninges or mid-brain, or via the alimentary tract to involve the pregnant uterus.

Clinical signs

The alimentary form often produces diarrhoea and even brief general illness, and if the ewe is pregnant abortion may follow a week or so later with retention of foetal membranes and subsequent systematic illness.

The much more common central nervous form appears to have a longer incubation period, cases not usually occurring within a month of silage feeding. The signs are variable but classically there are signs of unilateral V and or VII cranial nerve disturbance, which may show as a drooping ear, eyelid and lip with consequent dribbling and difficulty in eating and drinking. It also means that the animal is unable to respond to the menace test which may mislead one into thinking that the ewe is blind in that eye, and so confuse the condition with, for example, gid. However if the face and lip and gum are tapped or pinched there is often a loss of sensation on that side. The ewe also often shows a head aversion and a one-sided stiff neck, and she circles in one direction and may fall over onto one side. Usually, the ewe is noticeably very disturbed, ranging from the depressed to the convulsive. Deterioration usually occurs over just a few days, leading to recumbency and death.

Diagnosis

1. The alimentary/abortion form requires faeces, vaginal discharges, foetuses and milk (if any) for culture and paired blood samples for serology.

2. The CNS form requires differentiating clinically (if you can!) from such conditions as gid, CCN and pregnancy toxaemia, and the mid-brain examined histologically. Serology is not usually helpful.

3. If, as expected, silage is being fed, examine it for 'rotten' material and take samples for culture.

4. Be careful when handling the material (zoonosis).

Treatment

The organism is sensitive to a wide range of antibiotics but therapy is usually disappointing. It is worth trying, for example, high dose oxytetracycline (5 - 10ml Terramycin Q.R. 100 i/v followed by i/m every 12 hours for a few days).

Control

1. Silage:

Although a few incidents do occur at pasture, most follow silage feeding. It is almost impossible to avoid contamination by Listerella, so one must seek to avoid conditions in which it multiplies.

Suggestions:

(i) Make only high quality silage with a pH<5, with additives where necessary, and avoiding gross soil contamination e.g. mole hills.

(ii) Completely seal the silage, and so avoid air getting into the sides and tops of clamps and through insecure polythene (including bales).

(iii) Avoid feeding rotten silage, which can be seen and smelt on the top or sides, as a layer or even in lumps in the middle.

(iv) It seems less risky to feed silage to cattle.

2. Aborting ewes:

As always, isolate and clean up after (be careful).

3. Vaccination

Not yet available, but in the pipeline?

CEREBRO-CORTICAL NECROSIS (CCN)

(Polioencephalomalacia, 'brain rot')

This is an acute exciting central nervous disease, mainly affecting lambs 2 to 6 months old (not under 2, as it requires a functional rumen), but it does occur occasionally in adults. It usually occurs sporadically and 'out of the blue' but there are occasional outbreaks affecting a small number of sheep, and lasting a few weeks following a history of change in food or worm drenching. It is associated with a deficiency of thiamine probably induced by the microbial production of thiaminase in the rumen, causing an initial cerebral oedema followed by 'pressure necrosis'.

Clinical signs

After a short period (few hours) of aloofness, dullness and wandering (even circling!), the sheep becomes increasingly excitable (over another few hours even up to 2 days) developing tremors, staggering, recumbency, opisthotonus and galloping movements. It is usually blind (list the other causes of blindness). The sheep is often scouring (list the other causes of scouring). It may just be found dead (list the other causes of sudden death). The age of the sheep, the duration and convulsive nature of the disease may help to differentiate it from such conditions as pulpy kidney, pregnancy toxaemia, hypomagnesaemia, louping ill, listeriosis and gid, but it can be very difficult to avoid a blunderbuss approach at the first visit!

Diagnosis

(1) Clinical signs and response to thiamine treatment.

(2) PME

- (a) discoloration of cortical gyri and fluorescence under Wood's lamp of sliced cortex.
- (b) histology.
- (c) rumen contents for thiaminase estimation.

(3) Faeces for increased thiaminase activity, and bloods for increased pyruvate and decreased transketolase activity can give supportive information but they need special sampling procedures so contact the VI centre first.

Treatment

200 - 300mg thiamine e.g. 2ml Bimeda B1 or 5ml 'Parentrovite' (Beecham animal health) i/v (slowly) and i/m, repeat i/m daily for a few days. If the case has been 'caught early' i.e. before much irreversible necrosis, one expects some improvement within hours (not minutes), i.e. the sheep will be quieter and fewer convulsions, and often there is complete recovery within days (although vision often takes longer to return).

Control

It is reasonable to look at the diet and perhaps suggest changing (compare PK control), e.g. remove from that field or take off that concentrate or lick (e.g. molasses) but it is only making a shaky guess. Contacts might be injected with 500mg thiamine and perhaps given oral B1 and antibiotics, but the incidence does not usually justify such intervention. There is some evidence that the contacts do not thrive and show signs of scouring, so have a look at them too.

COENURIASIS (GID, STURDY, BENDRO)

This is one of the most common causes of disease of the central nervous system in sheep. The causative agent is the metacestode (larval) stage of the life-cycle of *Taenia multiceps*, (*Coenurus cerebralis*), a tapeworm occurring in the dog.

The adult cestode also occurs in the fox but the worms seldom reach maturity and become gravid. The metacestode is also recorded in cattle, goats, horses, deer and man, but at a much lower incidence than in sheep.

The larvae migrate in the bloodstream throughout the body but can only continue their development within the CNS. Over a period of 2 to 8 months, the Coenurus grows into a fluid-filled cyst containing around one hundred scoleces. Multiple cysts rarely occur in natural infections, possibly because an established cyst inhibits the development of further cysts.

Dogs usually acquire the infection either through scavenging dead sheep, or through being fed sheep heads obtained from abattoirs. Surveys have shown that about 5 to 10% of farmdogs in the British Isles are affected with the adult tapeworm, and that they carry a mean worm burden of about 10 adults.

Clinical signs

Acute coenuriasis usually goes unnoticed. A few days after a sheep has ingested a large number of the tapeworm eggs there is a brief period of pyrexia and listlessness, thought to be due to a diffuse inflammatory process resulting from toxic and allergic reactions to the parasites.

It is the chronic form of the disease that is more dramatic and therefore more frequently seen. The onset of clinical signs is usually 3 to 8 months after infection. The affected animal shows a variety of neurological deficits usually correlating closely with the location of the cyst within the CNS. About 80% of cysts are located in the cerebrum, 10% in the cerebellum, 5% multiple cysts in several locations, and the remainder in the brainstem and spinal cord.

Affected sheep are usually identified because of abnormal or sluggish behaviour. Neurological examination should then reveal specific localising signs although it is important to consider combinations of signs rather than relying too heavily on any one deficit.

Clinical sign		Interpretation
Unilateral blindness		Contralateral cerebrum
Postural deficits e.g. wheelbarrow, hopping tests		Contralateral cerebrum
Head aversion (head rotated and nose averted to either side)		Cerebrum, either side
Circling	- wide circles	Ipsilateral cerebrum
	- tight circles	Contralateral, lateral ventricle/ basal ganglia (some brainstem cysts cause circling)
Visual deficits		Occipital (caudal) cerebrum

Depression	Rostral cerebrum
Excitability	Temporal (lateral) cerebrum
Head tilt (head rotated but nose points forwards)	Vestibular or cerebello-vestibular region
Head tremor (especially intention tremor)	Cerebellum
Dysmetria (usually hypermetria)	Cerebellum
Nystagmus	Cerebellum

As the cysts develops the clinical signs gradually increase in extent and severity, progressing to recumbency and death (over days or weeks) if the cyst is not removed surgically.

Differential diagnosis

Coenuriasis usually occurs sporadically, although occasionally outbreaks do occur. The disease is characterised by the slowly progressive development of CNS signs in sheep over 4 months old (usually 1 to 2 years and rarely over 3).

Listeriosis is an acute disease (2 to 3 days), often with facial paralysis, occurring as mini-outbreaks mainly in winter, and usually associated with silage feeding.

CCN is also an acute disease (1 to 3 days), resulting in staggering gait, tremor and fits, but responding to early therapy with thiamine.

Louping-ill is a febrile illness characterised by 'leaping' movements, and usually occurring as outbreaks amongst yearlings in tick areas.

Other conditions to be considered include CNS abscess, swayback, intracranial injury, Border Disease, scrapie, pregnancy toxaemia, hypocalcaemia and hypomagnesaemia.

The chronic nature of the disease and age of affected animals, together with a known farm incidence, will aid diagnosis. In addition, the presence of skull softening at or near the horn bud site, due to the intracranial pressure, is diagnostic.

An intradermal test is sometimes used, but is unreliable because of false positives and negatives. The test involves the injection of 0.1 ml of cyst fluid (kept frozen from a previous case) into the skin of the neck or caudal fold. A positive result is indicated by a swelling after 30 minutes, as compared with a control injection of 0.1 ml of physiological saline.

Treatment

Lambs near killing weight should be sent for slaughter. Surgical removal of the cyst should be attempted in breeding animals or those too small for slaughter. With practice, the surgical success rate is high (about 80%).

General anaesthesia is obtained with i/v pentobarbitone (Sagatal - May & Baker) at a dose of approximately 1 ml per 5 kg body weight. Prior to surgery, crystalline penicillin (1.5 g) and dexamethasone (6 mg) should be given intravenously. A C-shaped skin incision is made just caudal to the horn bud on the appropriate side (L to R cerebral site), or across the midline just rostral to the nuchal line (cerebellum). The skill is trephined (0.5 - 1.5 cm diameter) and a circular plate of bone removed. Sometimes the skull is so soft that scissors may be used. However, it is not safe to assume that the site of skull softening indicates the location of the cyst. The meninges are cut and reflected to expose the bulging cerebrum (due to increased intracranial pressure). A 16 gauge needle and cannula (e.g. horse i/v catheter) is inserted. When the cyst is punctured, clear cyst fluid wells up the needle. The needle is removed and a 20 ml syringe is attached to the cannula so that most of the fluid can be withdrawn. Suction is used to trap the pale grey cyst wall in the end of the cannula which is then elevated to allow the pedicle of cyst wall to be grasped with artery forceps. Gentle tension is then applied until the wall is removed, draining more fluid from the cyst if required. Only the skin incision is sutured, and antibiotic cover is continued for 2 more days. The sutures are removed in 10 days and the skull heals in about 1 month.

The results of surgery can be very rewarding, the animal returning to apparent normality within 7 days in the majority of cases.

Control

1. Worm dogs with a drug against cestodes e.g. praziquantel ('Droncit', Bayer) every 3 months.

2. Do not feed uncooked sheep heads to dogs.

3. Dispose of sheep carcases properly.

4. Avoid sheep grazing heavily infected pasture e.g. following sheepdog trials or use by hounds.

SCRAPIE

This chronic progressive degenerative central nervous disease of adult sheep (and goats) appears to be widespread, as individual cases keep 'popping up' in flocks throughout the country. Usually just one or two cases occur from year to year in a flock (cf. Jaagsiekte and Maedi), and so the disease is of little commercial importance to the average commercial breeder; but for the pedigree breeder where older sheep may be retained and where there is a susceptible strain, the incidence may be significant and a conspiracy of silence often exists, with the breeder culling cases on suspicion and the true incidence never revealed. The causal agent has not been isolated, but it is probably transmitted in the lambing pens vertically (and perhaps horizontally) and some breed lines appear to have a built-in susceptibility, hence the culling of particular families and rams.

Clinical signs - probably depend on which strain of the agent is involved, and one strain may dominate in a flock and then only one clinical picture is evident.

(1) The most common and obvious sign is scratching, rubbing and nibbling in an adult sheep (usually over 3 years). It is often seen at first as restlessness with the sheep darting about from place to place (trough to trough) as if it has been bitten, and turning to nibble anywhere it can get

at, as well as rubbing on posts and walls etc. (watching a case one could imagine it feels like 'prickly heat'). If one rubs the sheep it responds by standing still and nibbling its lips in apparent pleasure and relief. (Note that this test is applicable for all conditions causing pruritis and is not diagnostic of scrapie).

The rubbing leads to a loss of fleece or hair over any area which the sheep can get at, including the head, face or lips. A fine rough stubble is left, often with scabs.

(2) Some show an obvious change in temperament but without obvious skin irritation, the sheep appearing wild and excitable, and with it an inco-ordination, so that the sheep becomes awkward to catch and to examine, and it won't relax and allow its head to be held comfortably, and swallowing is sometimes difficult. Rams can be quite dangerous and the difficulty in raising the head can suggest a high cervical lesion e.g. from fighting. Some other cases just look dazed and depressed and don't appear to focus properly.

(3) Over the weeks and sometimes months, all these signs worsen although there is sometimes a temporary stasis or even remission, all become progressively thinner and die. Some appear to worsen after resting e.g. overnight, but after being helped up they become more mobile and alert.

Treatment

The condition is always fatal, so one should cull as soon as possible.

Diagnosis

The clinical signs are eventually diagnostic but one obviously has to rule out primary skin conditions (e.g. sheep scab!) and all the other CNS disorders affecting adult sheep! Brain histology is required for confirmation; there are no serological tests.

Control

(1) Cull the affected, preferably before lambing.

(2) Fatten any offspring.

(3) Look at the family line including the ram and remove it, but only where the problem is sufficiently important.

(4) Lamb any suspects in isolation and destroy foetal membranes.

LAMENESS

GENERAL

Lameness is so common in most flocks of sheep that farmers perhaps erroneously, regard it as a 'fact of life' and give it only irregular attention. Apart from the discomfort to the sheep, the loss in production can be considerable, e.g. 0.5 kg/week in the fattening lamb, and inadequate food intake by the pregnant and lactating ewes contributing to pregnancy toxaemia and neo-natal diseases.

Foot rot in all its forms is generally the most important condition causing lameness in sheep, but examination of a number of lame sheep in a flock will usually reveal a number of other causes e.g.

(1) Interdigital problems caused by soil and grass 'balling'.
(2) Necrotic tracks and pus in the foot wall (as in cattle and pigs).
(3) Gross overgrowth leading to splitting and exposure of sensitive structures.

Although, again as in pigs and cattle, the foot is the most common site for causes of lameness, there are a number of important conditions affecting other parts of the limbs.

e.g.
(1) Injury - various
(2) Joint ill - tick pyaemia
(3) Erysipelothrix - (a) Joints; (b) Post-dipping
(4) Orf - Strawberry foot rot (see later)
(5) Muscular dystrophy (see later)

The prevalence of lameness in a flock varies greatly with climate, pasture, age and intensification e.g. over 50% of lowland lambs can be lame with foot rot if the weather and pasture conditions are suitable, whereas the prevalence in hill flocks is nearly always low; similarly, many lowland sheep may be lame with soil balling, a condition which rarely occurs in hill sheep.

This background incidence of lameness makes for problems in suspecting and diagnosing foot and mouth disease.

FOOT AND MOUTH (F & M) DISEASE

Fortunately, through lack of contact, this is likely to remain a rarity in sheep in the U.K. It seems unlikely that F & M will arise as a primary outbreak in sheep because they are unlikely to be fed imported unsterilized meat. This, therefore, means that practitioners will only be on the look out for F & M in sheep in their area, following known outbreaks in other species. F & M in sheep can present considerable problems, both because of the great difficulties in collecting and examining all the sheep in a flock and because sheep are known to act as carriers for several months.

Clinical findings

(1) Although the disease can be very mild in sheep, most people report systemic illness, with the sheep separated from the rest of the flock, looking unwell and unwilling to walk. (If you cannot catch the sheep it is not F & M ??).

(2) Lameness, usually in more than one leg, is the most obvious feature. Early lesions are seen as blanching and separation of the interdigital skin at the coronary bands; after a day or so the lesions are ulcerative and granulomatous and easily confused with more common causes, e.g. early foot rot.

(3) Mouth lesions are less common and less severe, and therefore, sheep may eat and usually do not dribble frothy saliva as in cattle. If present, blanching and separation of the mucous membrane occurs, most commonly at the dental pad, but again becoming ulcerated within a day or so of the initial blister.

(4) Ewes may abort and young lambs die (viraemia and myocarditis). If you are in any doubt, ring the boss or the Ministry (for a diagnostic visit), <u>and stay on the farm</u>. If there are other stock on the farm enquire about lameness and examine.

FOOT ROT

This is an infectious disease of the sheeps' foot, starting as a superficial interdigital dermatitis with subsequent separation of the horn. It causes much lameness and merits more vigorous attention than it often gets.

Cause

The synergistic activity of Bacteroides nodosus and Fusobacterium necrophorum invading softened or injured interdigital skin and further aggravated by secondary organisms such as spirochaetes and corynebacteria. There are many strains of B. nodosus and they vary in their invasiveness.

Clinical Picture

Two main types are traditionally described -

All types start in the interdigital skin, and show as a moist hairless raw area extending to both digits. The condition at this stage is similar to foul in the foot of cattle but without the obvious soft tissue swelling and degree of pain and is identical to scald. Often at least two feet (rarely all four cf. F & M) are involved at the same time, leading, for example, to walking on both knees ('praying') with consequent rubbing and loss of hair over carpuses. Sheep with foot rot can run about and it is often difficult to pick out individually affected sheep when a group are penned together, without turning all the sheep up.

The most virulent strains invade across the sole, SEPARATING insensitive structures and, if left unchecked, separation will continue up the walls to the coronary band, with consequent shedding of portions of the horn. This active 'separation', or under-running rather than destruction of tissues, facilitates paring, and as deep sensitive structures are rarely involved and secretory tissues are not irreversibly damaged, self-cure does occur. There is, however, little naturally induced immunity and, therefore re-infection is common, and feet may show more than one stage of the disease.

There is usually a soft cheese-like accumulation of debris between the layers, with the supposedly characteristic smell of necrosis produced by FUSIFORMIS organisms.

A condition affecting only the interdigital skin is called Ovine Interdigital Dermatitis (OID) or SCALD. This probably results from infection by F. necrophorum without the synergistic presence of B. nodosus. The lesion appears the same as the primary stage of the two forms of foot rot, but little horny separation follows. It sometimes occurs as outbreaks in young lambs grazing on wet or icy pasture, causing substantial lameness; it can also be a considerable problem in housed sheep if the straw yards are wet and warm. It rapidly responds to spraying or foot bathing, but is not controlled by the current foot rot vaccines.

Clinically, this all adds up to the following:

(1) If there is considerable horn involvement, then severe foot rot is present and paring is important, as well as other treatments.

(2) If the separation is confined to the heels in all cases, then a less invasive form of F. nodosus is present, and foot bathing will be more essential than foot paring.

(3) If the lesions are confined to the interdigital structures then Scald is present; spraying and foot bathing will be useful but more reliance will be placed on dry weather/bedding.

(4) If a mixture of lesions are discovered (this is the most common finding) then a mixture of strains and organisms are probably present and a mixture of treatment is necessary, and vaccination should be considered.

Some clinically important features of the 'Fusiformis' bacteria

F. necrophorum: Environmental organism with many hosts, therefore cannot be eradicated.

B. nodosus: Confined to sheep (and goats) and only viable for 3 weeks on the ground, therefore can be eradicated.

Require warmth and moisture (seasonal incidence)
Anaerobic (exposed by paring and try not to bandage)
Sensitive to chemicals (easily killed)
Smelly necrosis (diagnostic lesions)
Synergistic (useful to control only one)

Prevalence

This varies with rainfall and temperature, and also the concentration of sheep's feet. It is highest in warm wet weather when there are many ewes and lambs together i.e. lowland flocks in spring and autumn, and it can be artificially induced by wet bedding indoors (housed sheep). It is probably lowest in hill sheep where concentration is least and temperatures low (plus the additional factor that the acid peat soil may not support Fusiformis).

Treatment

(1) **Individual lame sheep:**
Careful paring of any separated horn and spray with a tetracycline aerosol.

(2) **The flock:**
Walk slowly through a well-designed foot bath and keep on the concrete draining pens for 15 min. Use zinc sulphate (Golden Hoof - Sheep-Fair Products Ltd.). Repeat once a week at times of peak incidence. These chemicals are probably protective for 1 week and it is, therefore, not essential for sheep to go on to "clean" pasture after foot bathing ("clean" pasture = pasture not grazed by sheep for 3 weeks).
*A new preparation 'Footrite' (C. Vet. Ltd.) containing zinc sulphate and a penetrating agent, claims to be very effective if sheep are held in a suitable footbath for an hour. It is initially expensive but could lead to eradication; it merits evaluation.

(3) **Severe cases and very valuable stock (e.g. rams, heavily pregnant ewes):**
One large dose of penicillin e.g. 10 ml and streptomycin i/m plus paring and foot bathing.

Control

(1) Most shepherds routinely trim and foot bath, but there is often no planned control and sheep and the shepherd learn to live with foot rot. The more energetic the programme the more effective is the control.

(2) **Vaccination**
There are two vaccines now available - 'Clovax' - Coopers, costing approx. 20p/dose and "Footvax" - Coopers, costing approx. 40p/dose and containing more strains of B. nodosus. Both require 2 doses initially and boosted twice a year, according to a prescribed schedule e.g. February, but NOT just before lambing, and October. However, there is a need for maximum protection during housing and if this coincides with late

pregnancy it may be wise to vaccinate at housing and again shortly after lambing. Where the incidence of foot rot is obvious, vaccination should be seriously considered both for the sake of the sheep and the pocket.

'Footvax' produces a noticeable lump, which usually recedes over a few weeks.

Eradication

Foot rot has been eradicated in some flocks but it is a difficult and time-consuming exercise and by no means certain. It certainly requires that the farmer be prepared to examine all feet, effective pare and foot bath, repeating the procedures at least once, and isolate or even cull persistent cases. Eradication implies safeguards against re-introduction and this means self-contained flocks (including adequate sheep fencing) or inspection and eradication of infection in purchased sheep. Few flocks, particularly the large ones, can manage all this. It is best to attempt it in late summer when there are few lambs about.

Foot Abscess

A variety of lesions occur which affect the wall of the foot, similar to those found in pigs and cattle. The lesions are usually a local necrotic laminitis with secondary abscessation and eventual bursting at the coronary band. Usually only one digit is involved, which is very painful when squeezed. Treat initially by paring and intensive systemic antibiotic therapy, i.e. high dose, twice daily for 1 week; the occasional very severe and valuable cases e.g. rams, many justify amputation, which is relatively cheap and easy. X-rays help to decide if the joint (P.2/P.3) is involved; as does a probe and the degree of swelling and lameness (usually 10/10). but even some of these cases slowly resolve without amputation.

Interdigital 'Fibromas'

Proud, fibrous outgrowths in the interdigital space (mainly hind feet) is a feature of some breeds e.g. Suffolk rams are notorious. Lameness occurs only when they become ulcerated and infected. Frequent foot-bathing maintains some control, but the worst cases require simple surgical removal under local anaesthesia; this is not a job to be done just before tupping!

JOINT INFECTIONS

Young lambs up to 1 month - mainly lowland

Following a septicaemia of usually streptococci or Escherichia coli via navel or gut, there is a polysynovitis and arthritis, particularly of the stifle (soft, bulging and painful synovia in the patella region) carpus and hock joints, causing a reluctance to move about, with flexed joints and arched back (crouched and crippled). It is best to assume an E. coli infection and treat accordingly and vigorously. Standard methods of control for E. coli are indicated, but it is not always easy, nor it is always obvious what it at fault. "Coliovac" - Hoechst, a polyvalent E. coli vaccine is worth considering in future if the incidence justifies the expense.

Lambs over 2 months - lowland (Stiff lambs/Erysipelas arthritis; Chlamydial polyarthritis)

On some farms a significant number of fattening lambs at grass show a stiff, stilted, short-striding walk or hopping run, appearing to be lame on more than one leg. Very careful examination and palpation discloses some soft synovial swelling and pain in some limb joints, particularly stifle, hock and carpus. The condition is insidious, and often irreversible by the time it is recognised. Erysipelothrix is the 'favourite' organism causing a chronic fibrinous synovitis, often with no or only a little cartilaginous involvement and nothing to show on X-ray. A blood titre of 1/320 and over indicates infection with this organism. Penicillin, preferably containing a long-acting component, is probably the drug of choice and it is tempting to use corticosteroids and analgesics e.g. Leucotropin. Vaccination (e.g. Erysorb - Hoechst), may be indicated, including vaccinating ewes in late pregnancy (as for clostridia). It is perhaps useful to try and establish whether there is a particular field, season or age group which becomes affected on the farm, and apply measures in this group only. Such lame sheep are worth housing so that more intensive treatment and feeding can be applied, and where the lambs do not have to walk far for food or shelter. Repeated penicillin treatment is advisable.

Erysipelothrix can also invade broken skin. At the time of dipping, the organism from the soil can seriously contaminate the dip and enter limb wounds caused by handling at this time; within a day or so, many sheep may be very lame with a local cellulitis of the hairy parts of the limbs - POST DIPPING LAMENESS. Early treatment with penicillin is essential and in future, add a bacteriostat to the dip (usually incorporated now, e.g. 1:1000 Zn SO_4).

Lambs up to 4 months - hill

Following tick bites and staphylococcal septicaemia (see later).

LAMBING TIME

HOUSING EWES

Advantages

(1) Housing prevents damage to the pasture by the sheep, by poaching of land during the wet winter and by the nibbling of young grass shoots in February and March. This is particularly important on wet, heavy land and allows good grass growth in April when ewes are lactating heavily. This allows an increase in stocking rate which must follow housing if the costs are to be covered. Housing usually commences in December/January and sheep are turned out near the end of March, depending on weather.

(2) Housing improves the working conditions of the shepherd and protects sheep (and new-born lambs) from extreme weather conditions. Ewes are more carefully observed and shepherded during lambing which reduces lambing losses.

(3) Housing allows ewes to be sorted into different batches according to condition score, number of lambs being carried (if ultra sound pregnancy detection is used), state of teeth, lambing dates etc. and feed adjusted accordingly.

(4) Housing reduces the wastage of food if well-designed racks are used, whereas considerable wastage of concentrate occurs in adverse weather if fed outside. The feeding of the ewes is more controllable and regular condition scoring allows appropriate adjustment to feeding to be made.

(5) Housing allows (and usually necessitates) the shearing of ewes (see later).

Disadvantages

(1) Capital costs: These depend on the type of building and whether existing buildings can be modified. The building does not need to be of complex design but a purpose designed sheep house could cost up to £40 per ewe with an additional £5 for troughs. Much cheaper buildings are available, the cheapest being the 'polythene tunnel'. The latter building, composed of steel hoops covered with polythene and usually with walls of plastic mesh

costs about £10 - £15 per ewe. The polythene covering has to be replaced periodically, some types every 3 or 4 years, but some last 10 years. Grants are available towards the cost of housing, ranging from 22 1/2% to 50% in less favoured areas. Most buildings may be used for some other venture (e.g. calves) during the summer and autumn.

(2) Disease risks: Intensive housing may increase the risk of certain diseases such as neo-natal scours, coccidiosis, foot rot and pneumonia but these can be controlled by good construction and/or management.

(3) Food and bedding costs: Food costs in March will be increased because the young grass is deliberately not available to the ewes but this will be beneficial in the long run. Straw is an additional cost.

Principles of Design

Site

The house should be convenient for the shepherd, be sheltered from the prevailing wind and snow and have ready access to water and electricity.

Design

It is usual to have a wide centre passage, which may hold hay and concentrate feed and be used for lambing pens when the hay has been reduced. Pens to hold 30 - 60 ewes are sited on either side of the passage. The centre passage is best concreted but hardcore is best for the pens to permit good drainage. The size of pens varies with breeds and shearing but is about 1.0 - 1.3 sq. metres (10 - 14 sq. ft.) per ewe. It is most important that the ewes have ample trough space (0.45m, 18 inches each). A useful design is to have removable, double-sided pen dividers consisting of a lower food trough and upper hay rack with a walk way along the middle of the rack. Straw bedding should be provided or slatted floors. About 2 bales of straw are needed for each ewe for a 3 month period. The house must be well ventilated above the sheep by Yorkshire boarding, plastic mesh, or open sides but be draught-proof at sheep level by having solid walls to a height of 1.0 -1.2m. Water may be provided in troughs, self-fill bowls or continuous flow along rainwater gutter. The water should be at a height that the sheep can reach but not too low or it will become contaminated with faeces and thought should be given to the increase in bedding height which occurs during housing. Good lighting is essential; artificial light should be 4 watts/sq. metre of floor space. Power points are also essential, for shearing and particularly for infra red lamps over the individual lambing pens, and for the shepherd's kettle!

[Information on housing design and use of slats may be obtained from: Farm Buildings Information Centre, National Agriculture Centre, Stoneleigh, Kenilworth, Warwicks. CV8 2LG.]

SHEARING HOUSED EWES

Advantages

(1) It promotes a drier and cooler environment, with less heat stress in the ewes which therefore appear more comfortable.

(2) There are fewer disease problems associated with humid conditions e.g. pneumonia, foot rot and lumpy wool.

(3) Ewes take less space, including trough space and more can be kept in a pen (up to 20%).

(4) Ewes eat more (if it is supplied!).

(5) Birthweight of lambs is increased and there are fewer cases of pregnancy toxaemia (if (4) applied).

(6) Thin ewes can be spotted easily and action taken (but very thin ewes should not be shorn).

(7) Ewes are cleaner and 'hung' lambs easier to see, promoting better lambing conditions for ewe, lambs and shepherd.

(8) The quality of shorn fleece improved.

(9) There are fewer problems with lice and keds.

(10) They are easier to blood sample.

Disadvantages

(1) Good housing is necessary if the sheep are not to become too cold and huddle together.

(2) Sometimes turn-out in the spring has to be delayed because of bad weather, leading to problems with housed lambs, bedding, feet and food costs. (At least 2 months' wool growth should be allowed before turn-out).

(3) They are sometimes too hot in the summer with six month fleece.

(4) There are increased feed costs.

(5) Single lambs may be too large, causing dystocia.

(6) 'Wool-Slip' - some wool loss is often noticeable in a proportion of the ewes. Various reasons have been suggested but the most likely appears to be that the dual stress of housing immediately followed by shearing causes increased blood steroid concentrations, which is known to cause wool loss.

(7) The skin is more easily injured and infections such as ringworm and staphs. become more widespread.

(8) There is a loss of quantity of fleece in the first year.

(9) Ewes more difficult to catch and to truss.

CLOSTRIDIAL DISEASES

Clostridial diseases, other than tetanus, generally cause rapid death, therefore the practitioner is really only concerned with the problems associated with diagnosis and control.

CLOSTRIDIAL INFECTIONS

DISEASE	CLOSTRIDIUM SP	AGE	SEASON	TRIGGER FACTORS
1. Lamb dysentery	perfringens B	< 2 weeks	Spring	flush of milk
2. Struck	perfringens C	< 2 weeks and adults	Spring	flush of milk or grass
3. Pulpy kidney	perfringens D	> 2 weeks	any	flush of milk or grass or concentrates
4. Tetanus	tetani	2 - 4 weeks	Spring	docking/castrating
5. Braxy	septicum	4 - 8 months	Autumn	frosted food
6. Black	oedematiens B	adults	Winter	fluke
7. Bacillary haemoglobinuria	oedematiens D	adults	Winter	fluke
8. Blackleg Post-Partum gas gangrene Malignant oedema	chauvoei septicum	any	any	injury/wounds

Diagnosis

This needs to be speedy if control is to be effective. It is usually dependent upon obtaining from the farmer, accurate and relevant historical information (one is looking for a plausible excuse for the disease) supported where possible by post-mortem examination (PME) of fresh carcases. It is important not to rely on PME alone, not least because the material is often too decomposed for useful interpretation.

The information you need is:-

(1) Did the sheep die 'suddenly'? (found dead)

(2) Were there any signs? (scouring, convulsions, stiffness)

(3) What age? (note the age ranges for the different clostridial diseases)

(4) What has the farmer done to control clostridial diseases? Have vaccines and/or antisera been used? (If so, closely check what and when)

(5) What has the farmer done to excite or introduce the disease? (e.g. change to better pasture or more concentrate, inject with dirty needles, used rubber rings).

(6) Has the disease occurred on the farm before? (e.g. Lamb dysentery, Black)

(7) What season is it? (spring grass excites pulpy kidney and struck, fluke in winter excites Black, frost in autumn excites braxy)

(8) Are the affected sheep in good condition? (clostridial diseases often 'select' the greedy, fatter animals)

(9) Post mortem examination.

Control

Clostridial vaccines are so potent, so cheap and so available, it is not unreasonable to expect very few losses from clostridial diseases. However, V.I. centre reports continue to show that serious losses still occur, and that these are usually due to mistakes in application e.g. choosing the wrong mixture, injecting at wrong times, omitting injections and 'dirty' injections. The Vet has a significant part to play in this area.

The recipe for control contains three ingredients:-

(1) Avoid the exciting factors
In practice this really means the control of fascioliasis, the use of sterile instruments and working in clean dry conditions (including the sheep).

(2) Antibiotic cover (e.g. long-acting antibiotic) following trauma e.g. castration and assisted lambing.

(3) Antibodies - derived via vaccination, colostrum or antisera.

Vaccination: There are a number of vaccines (e.g. Covexin 8, Tasvax 8) which protect against all the important clostridia. By using an 8 in 1 vaccine and efficient primary and secondary doses, it is possible to maintain a protective level of immunity throughout the year in both ewes and lambs, against all the usual clostridial diseases. Some other vaccines contain fewer antigens and may be cheaper, but it is arguably better to have a fully comprehensive insurance policy for these killing diseases. However, 7 in 1 vaccines are much used (omitting Cl. oedematiens D) and 'Heptavac P' - Hoechst, clostridial protection is combined with pasteurella, and the schedule for vaccination fits both nicely. In Nilvax - Coopers, an 8 in 1 clostridial vaccine is combined with the anthelmintic levamisole - arguably an expensive way of dealing with the problem.

Two doses of vaccine costing approximately 15p/dose, at 6 weeks' interval produces sufficient antibodies to protect a ewe for one year with sufficient spare, via colostrum to protect its lamb(s) for up to 16 weeks, providing the secondary or booster dose is given approximately 1 month before lambing.

Assuming a flock to have no initial protection, one effective vaccination schedule us:-

(a) Ewes and Rams

(i) Primary
(ii) Secondary - 6 weeks later
(iii) Booster - 2-4 weeks before lambing
(iv) Repeat (iii) annually

(b) Lambs

Those to be kept over 16 weeks (for fattening or breeding)

(i) Primary - at 12 weeks
(ii) Secondary - at 18 weeks
(iii) Booster (for breeding lambs only) at flock pre-lambing time (along with the adult ewes)

Careful injection technique is needed to avoid abscess formation and trimming of carcases (use neck site and s/c).

(c) Bought in ewes (for tupping) and lambs (for fattening)

It is usually best to assume that these have not been vaccinated, or at least their immunity has waned and to inject them 2 or 3 times in the first winter. A short-cut is often taken in the bought-in ewes (with some justification if there is no obvious winter clostridial risks e.g. Black disease), and that is to vaccinate them on arrival but delay the secondary injection until 2 - 4 weeks before lambing.

(d) Lambs out of unvaccinated or inadequately vaccinated ewes:-

EITHER: (i) Primary - in first week, together with Lamb Dysentery (LD) antisera if LD is a known risk.

(ii) Secondary - at 6 weeks.

OR Rely on 100ml of colostrum (fresh or frozen) taken from a vaccinated ewe (or cow) and given at birth.

It is important to note that apart from avoidable mistakes that farmers make, it sometimes proves to be particularly inconvenient, and sometimes impossible, to inject sheep at the right times (particularly the pre-lambing booster), and also that not all lambs, particularly multiples, obtain adequate colostrum. The Vet should be aware of such events on his clients' farms, and advise accordingly, in particular, ensure an adequate supply of frozen colostrum pre-lambing.

Cow colostrum is now very fashionable as a replacement for ewe colostrum. If good quality material is used (the first milking and before the calf is allowed to suck) it is clearly an adequate substitute food and most lambs tolerate it. (An anti-ovine factor has been discovered in some cow colostrum causing a disturbance in the lamb's bone marrow and subsequent anaemia; if this is suspected avoid this donor cow or her colostrum.) To provide clostridial protection with cow colostrum, current work suggests that the cow can be vaccinated with sheep 8:1 vaccine (10ml 3 months before calving, and again at 1 month and 2 weeks before calving) and 100ml of the subsequent colostrum (given by syringe or stomach tube) provides protection to the lamb for at least 3 months.

IN AN OUTBREAK, control measures in the rest of the flock consist of:-

(a) Remove from the exciting source (where possible) e.g. take off the lush grass or reduce the concentrates.

(b) Antibiotic injection (e.g. penicillin) for those with local sepsis e.g. 'gas gangrene' and tetanus.

(c) Antitoxin to all at risk, e.g. Lamb Dysentery, Pulpy Kidney, Tetanus.

As the immediate protection provided by antitoxin persists for only approximately 2 weeks, it is often reasonable to inject a primary dose of vaccine at the same time (but at a different site), followed by the secondary dose of vaccine 6 weeks later, thus providing both immediate and long-term cover.

ABORTION

General Incidence

The incidence of abortions in 'normal' flocks is usually quite low (1 - 2%) and is usually tolerated by the farmer without investigation. However, it must be remembered that some or many of these abortions are caused by pathogens or nutritional disturbances, which also cause apparently 'barren' ewes, foetal mummies and very weak non-viable premature lambs. Abortions, therefore, should always be considered as only part of the losses in the production process and must be used as markers of what is often a much wider disease problem.

Example:- In a 'normal' commercial lowland flock of 100 ewes, anticipating 150 live lambs, it is often true that up to 4 ewes will not produce lambs, 2 ewes will abort, and 20 lambs will be born either dead or die within a few hours of birth.

This means that the 100 ewes have only produced about 130 live lambs, and that about 10 of the ewes are 'unproductive'. In addition, probably 5 ewes will have died around lambing time for reasons not directly associated with abortions.

When, however, 'uterine' infection is introduced into a fully susceptible flock, the incidence of abortions is often alarming and is recognised as a 'storm'.

Causes

All abortions should be regarded as infectious until proved otherwise. The four most common causes of infectious abortion are:-

(1) Enzootic (Chlamydial) (EAE)
(2) Toxoplasma
(3) Vibrio (Campylobacter)
(4) Salmonella

A number of other agents cause sporadic abortions:-

Border Disease Virus
Listerella
Tick borne fever - Rickettsia
Q fever - Coxiella
Brucella
Fungi

The confirmed prevalence of these infections varies with the area of the country and perhaps the interest of the laboratory.

The summary diagnoses recorded at Veterinary Investigation Centres in England, Scotland and Wales for the years 1975-1982 inclusive give the following incidents of abortions in sheep (each a separate 'outbreak').

	NUMBER	%
Total Abortion incidents	21268	100
Diagnosis not reached	10679	50.2
Chlamydia	4600	21.6
Toxoplasma	3406	16.0
Campylobacter (Vibrio)	1095	5.1
Salmonellae	367	1.7
Other diagnoses	1121	5.3

Note the high % of incidents of abortions that are 'not diagnosed'. The most likely explanations for this are that laboratories:

(1) often receive useless material

(2) can generally only consider infectious agents (and then perhaps only a few of these)

(3) receive just the 'odd' abortion from the 'normal' flock.

As a general rule, if there is a serious abortion problem and the laboratory receives sufficient useful material, the cause of abortion becomes apparent.

The diagnostic laboratory requires:

(1) Fresh foetal membranes with cotyledons. (If these are not available then supply swabs of vaginal discharge.)

(2) Fresh foetus(es).

(3) Information which will include

Size and nature of flock e.g. is it self-contained?
Feeding of flock
Quality of shepherd/farmer
Previous abortion history
Dates of lambing
Facilities for isolation
Preparedness of therapy
Clinical excuses for the abortion(s), e.g. handling and dosing, predisposing diseases e.g. fluke and pregnancy toxaemia, and dogs (though the evidence for the latter is very slight!)

(4) The laboratory may also subsequently request a number of paired blood samples, and/or rectal swabs, so it is worth tagging aborting ewes so that they can be identified subsequently.

Remember that some ewes can produce twins, one of which is apparently normal, and the other abnormal and infected. A flock may be infected with more than one pathogen, e.g. Toxoplasma and enzootic, so that it is worth continuing to collect aborted material even after the first positive diagnosis has been established.

General Advice

Isolate aborting ewe and retain foetus(es) and membranes; treat as infectious (use poly glove and bags), not forgetting the public health risks from:-

Chlamydia
Toxoplasma
Salmonella
Listerella
Brucella
Q. Fever

This means that particular precautions should be taken for pregnant women and it is preferable that they should not be involved with aborting ewes, or perhaps, even with the lambing flock. Keep in isolation until a positive diagnosis has been made and usually until obvious vaginal discharge has ceased (up to 3 weeks) and tag so they may be identified later, if necessary, for blood or culling. If the ewe is ill, antibiotic therapy will be necessary after the vaginal material has been obtained.

Do not use aborting ewes or ewes producing premature lambs as foster mothers until it is known to be safe to do so.

It is usually better to retain aborting ewes and not sell them, because you hope that they will be now immune to at least one pathogen, and not abort again.

The laboratory findings e.g. Campylobacter may suggest that it is a good plan to allow the aborted carrier ewe to mix with the ewes and lambs which have already lambed, and so induce general flock immunity, i.e. 'move on' rather than 'move back'. It is essential, however, to be absolutely certain that EAE is not present also before recommending this practice.

Bought-in sheep should mix with the resident sheep, general farm environment and on-the-farm food for as long as possible (3 months) before tupping, but they should, where possible, lamb separately in their first season on the farm, in case they are infected with a 'strange' pathogen, and perhaps abort.

An abortion storm is usually not repeated in subsequent seasons and subsequent fertility is good, unless a new pathogen is introduced.

At the time of abortions, those ewes yet to lamb should be kept apart from the infected group and spread out, and the possibility of EAE vaccination and/or antibiotic therapy should be considered. Where possible, they should not lamb in the same area as the infected group, or at least the lambing yards should be re-strawed.

Consideration should be given to maintaining a closed flock, and to future EAE vaccination.

Enzootic Abortion (EAE)

General

Caused by Chlamydia psittaci (ovis) a bacterium which parasitises specialist host cells, forming elementary bodies. In some ways, chlamydial abortions present a similar clinical picture to other abortions viz.

(1) No systemic disease in either ewes or lambs, other than abortion and a few sick ewes following metritis.

(2) Following abortion, immunity is strong.

(3) The organism rarely causes disease in other hosts (although similar organisms are carried by other animals and birds and therefore there is always a risk of the disease pattern changing, and inter-species transmission may occur.

However, there are certain specific differences which make the clinical story quite different.

(1) The most important source of infection is from aborting ewes since the organism is mainly excreted at the time of abortion and in subsequent vaginal discharges, which resolve in three weeks. (An intestinal infection with faecal excretion of organisms, has been found on infected farms throughout the year. Recent work suggests that a small proportion of these faecal organisms are capable of infecting the placenta and leading to abortion but the full significance of this in the epidemiology of enzootic abortion is not yet known. It may be useful to know that a high

proportion of lambs on farms where enzootic abortion is present shed chlamydia during the summer and the organism can be isolated from these lambs to investigate an abortion outbreak retrospectively, along with serology of the ewes.) Infection is mainly spread at lambing time and therefore both the rate of infection and the rate of development of flock immunity is usually slow. Few 'storms' occur but the incidence of abortions tends to persist unless otherwise controlled. The disease is almost unknown in hill flocks where the management at lambing is less intensive than lowland, whereas it represents a big hazard for sheep lambing under intensive conditions e.g. housed.

(2) Lambs can be infected at birth or as yearlings, and often produce infected lambs and membranes at their first lambing and then become solidly immune. (i.e. infection of non-pregnant animals produces latent carriers rather than immunity, compare toxoplasmosis and vibriosis.)

(3) Ewes infected for the first time in late pregnancy do not usually abort then but may do next time. It takes about 40-50 days from infection to abortion, but ewes with widely differing lambing dates should not be mixed, as infection and abortion can occur within the same lambing season.

(4) The usually less violent clinical picture is further illustrated by the infection often showing as late abortions merging with still-births which merge with weak 'soft' live lambs which merge with viable lambs plus infected membranes.

Characteristically, in a non-vaccinated flock purchasing only a few replacements each year, the first season of infection shows itself as a few purchased ewes producing premature lambs; the next season abortions and still-births occur in the older ewes and the following seasons abortions etc. are mainly confined again to the yearlings and bought-in sheep.

Diagnosis

Where possible the laboratory requires foetal membranes, although the organism can be isolated from vaginal discharges and foetal lung and liver. Modified Ziehl-Neelsen staining of smears from the infected cotyledons show intracellular inclusion bodies. There is often an obvious placentitis with thickening and necrosis, so look at them! However, infection may not be obvious in the placenta when organisms are few and tissue culture will reveal organisms where smears have been negative. Examination of sera for complement-fixing (CF) antibodies where a positive titre of 32 or above can be useful, but it is not usually necessary unless membrane material is not available. Interpretation of CF titres must take account of any known vaccination history, and previous abortions because these may produce persistent titres. Paired samples 2 weeks apart provide the best information.

Treatment and Control

Aborting ewes rarely require any treatment other than isolation until their discharges cease (3 weeks). They should not be used as foster mothers but they should be retained. It is wrong to mix these with lambed ewes until discharges cease since the lambed ewes may become latently infected and abort next year.

It is unusual for the ewes yet to lamb to receive any treatment other than to try and keep them away from the infected lambing pens, where the organism is viable for approximately 6 weeks. However, when the risks appear to be high and/or the client is anxious to do more than 'sweat it out', a single injection of long acting oxtetracycline (dose for 80 kg ewe, 8 ml costing about £1.00), to those ewes likely to lamb in the next few weeks, is worth trying. In the first year after an outbreak in a valuable flock it may be worth injecting oxytetracycline 3 and 6 weeks prior to lambing to reduce abortions in ewes which were already infected.

Vaccine: There is an inactivated vaccine, Ovine Enzootic Abortion Vaccine (Coopers) costing approx. £2.25 per dose. It produces a more or less life-long immunity (3 years) and on the first occasion, all the breeding sheep are vaccinated in the late summer prior to, or at, tupping. In following seasons, only the replacement sheep require vaccination. Vaccinating ewes and gimmers which had been orally infected the previous lambing season may prevent uterine infection, although it will not be 100%, and therefore the farmer should be warned that some of them might abort or produce weak lambs. In an emergency, the vaccine can be given to pregnant ewes exposed to infection in an attempt to prevent further abortions and/or infections but it takes two months for neutralising antibody to develop, so the vaccine is only likely to have 'curative' value in those ewes more than two months off lambing.

In recent years there have been abortions caused by Chlamydia in ewes on farms where a proper vaccination programme has been used for many years and strains of Chlamydia have been isolated which are antigenically different from the vaccine (A22) strain. Recent work on a large number of isolates from all over Britain has shown that the A22 strain would protect against about 70% of these isolates.

The vaccine has been modified to include a second strain which should extend its cover but this is unlikely to result in complete cross protection to all the different antigenic types. Recent work indicates that this vaccine may prevent abortion but infected membranes still occur and infection is thus perpetuated on the farm.

Toxoplasmosis

General

Caused by the protozoan Toxoplasma gondii.

(1) It infects almost any host but it had been shown that the life-cycle can only be completed in the cat (and wild Felidae) in which sexual (as well as asexual) multiplication occurs in the epithelial cells of the intestine. This results in the production of large numbers of Isospora type oocysts which sporulate after a few days. Sheep and other hosts, including man, can become infected by ingestion of the oocysts, probably, in the case of sheep, from contamination of home-grown cereals, hay or bedding with cat faeces. The oocysts are highly infective and remain viable for months and susceptible sheep will show virtually 100% abortion if infected in mid-term, with as few as 2000 oocysts. Material containing cat faeces may be spread on pastures and sheep be infected whilst grazing.

Infection of pregnant women may result in foetal hydrocephalus and abortion.

(2) Infection early in pregnancy is likely to cause foetal resorption and subsequent return to the ram, whereas infection late in pregnancy (day 120+) will usually result in the birth of a normal lamb which may be infected and become immune. Infection in mid-term (days 60-120) will cause foetal death, mummification and abortion; the time from infection to abortion being about 40 days, similar to chlamydia.

(3) Immunity following infection is strong and ewes never abort more than once. If infection occurs outside the dangerous period, whether the ewe is pregnant or not, the ewe will become immune.

(4) Although it is known that rams can excrete the organism in their semen for a short time after infection, infection of the ewe at the time of tupping would not result in abortion.

(5) The suggestion that ewes may transmit infection to each other, either before or after aborting, is not borne out by recent work, which strongly suggests that the only source of infection for sheep is oocysts passed by cats.

(6) The clinical picture is usually a number of ewes aborting in late pregnancy (it may appear as a storm), with mummification of one or more foetuses, some weak lambs may be born and there is often a number of apparently barren ewes at the end of lambing. The ewes are not ill.

Diagnosis

(1) Fresh foetal cotyledons often show small (2mm) white necrotic calcified foci - 'White Spot Placenta'or 'Frosted Strawberries' and often attached to brown, dry mummifying foetuses. The white foci are often easier to see if a microscope slide is placed on the cotyledon, or the cotyledon pressed against the side of a clear poly bag. Smears of cotyledons stained by Giemsa or Leishman may show schizonts but they are very sparse.

(2) Histology of placenta and foetal brain may show Toxoplasma or specific cell reaction (foetal brain squash for pseudocysts).

(3) Transmission of ground portions of placenta and foetus into mice. This takes up to eight weeks, and, therefore, is only of use retrospectively.

(4) Bloods (10): The indirect haemagglutination test is valuable, high titres (> 320) usually indicating active infection, also repeated serological DYE tests at 1 month intervals. The titres can be persistent (because the infection is persistent) and therefore careful interpretation is required. A latex agglutination kit is commercially available for antibody measurement from Mast Laboratories, Bootle, Liverpool.

(5) Immunofluorescent techniques, e.g. fluorescent antibody on sections or smears of cotyledons.

(6) Serology of still born lambs of foetal fluids, precolostral live lambs or IgM antibody in postcolostral live lambs. The presence of antibody in the foetus indicates an active infection in the uterus at the dangerous time.

Treatment and Control

(1) During an outbreak of toxoplasmosis, aborting ewes are not dangerous to other ewes, or even useful, so there is no obvious justification in continuing to isolate them once the diagnosis has been established. However, the farmer can easily become confused with this exception to the rule, and one must also always be aware of mixed infections, and zoonotic risks (handle abortions with care with disposable gloves).

(2) During the outbreak there is usually little to be done other than to grin and bear it, although it is prudent to look for sources of cat faeces, and postpone feeding them until after lambing! Keep cats away from cereals, hay and bedding likely to be available to pregnant sheep (pelleted concentrate is likely to be safe).

(3) The organism is, however, susceptible to sulphonamides and injections (even one-off) of high dose trimethoprim and sulphadiazine (e.g. Tribrissen injection 48% at 60mg/kg, costing approx. £2.00 per ewe) is worth considering. Unfortunately, pyrimethamine (Daraprim), a preparation used in man, cats and dogs for acute toxoplasmosis, is not easily absorbed via the rumen, and is not useful for treating infected sheep.

(4) After lambing, retain the aborters and maintain within a closed flock, or mix bought-in replacements with resident sheep as well as the feed and farm environment, for as long as possible before tupping so that they are exposed to material likely to contain cat oocysts.

(5) It would appear to be a good disease for vaccination, so watch out! It has also been shown recently that giving a coccidiostat to the ewes will prevent oocyst infection resulting in abortion but more work is needed on this before practical application.

Vibriosis (Campylobacteriosis)

Although the organism causing this form of sheep abortion was originally classified as *Vibrio foetus*, it is now agreed that it belongs to the genus *Campylobacter*. There are two main classifications of *C. foetus* into subspecies so different names will be found in different publications.

Sheep abortion is caused by *Campylobacter foetus* subspecies *intestinalis* and *jejuni* (not *C. foetus* subs. *foetus* as in cattle).

(1) Infection is ingested, not veneral (c.f. *C. foetus* subs. *foetus* in cattle), and is primarily intestinal. It is therefore excreted in the faeces from symptomless carriers.

(2) Ewes are rarely ill at the time of abortion, but *C. foetus jejuni* can cause gastro-enteritis in lambs, calves, dogs and man (so be careful).

(3) Birds, e.g. crows and magpies, may introduce the infection to a flock as well as introduced carrier sheep, and the infection is then spread from sheep to sheep via faeces, abortions and personnel.

(4) Early foetal loss is rare as ewes appear to be resistant in the first 3 months of pregnancy; campylobacter infection is therefore not associated with 'barren ewes'. Abortions appear in late pregnancy 7 to 25 days after infection.

(5) Immunity following infection is strong, and subsequent fertility is good. Outbreaks may re-occur every few years because of lack of immunity in replacement stock, but usually abortions are few in number after the initial year of introduction.

Diagnosis

By smear and culture of cotyledons, foetal stomach or liver and visible focal necrosis of foetal liver is sometimes present. Campylobacter is not difficult to culture, but exact typing is complicated. Serology is not very helpful.

Treatment and Control

Usually the individual aborting ewe does not require any therapy. Once the diagnosis is confirmed, such ewes should mix with the ewes which have already lambed, but not with the pregnant sheep.

The remaining pregnant sheep must be kept away from the infection as far as possible and spread out, although the number and spread of abortions may not make this possible. Where a 'storm' is apparent, consideration should be given to antibiotic treatment of groups of pregnant ewes, with e.g. penicillin and streptomycin.

The farmer should consider keeping a closed flock, although infection can be introduced by birds. He should not sell the aborting ewes. If he does buy replacements, he should mix them with the resident flock for as long as possible before mid-pregnancy, but separate them in late pregnancy.

There is no vaccine available in the United Kingdom at present, although, elsewhere, a multivalent vaccine is known to be useful both during an outbreak and subsequently.

Salmonellosis

General

One host-specific salmonella S. abortus ovis, and also the less specific salmonella, e.g. S. typhimurium, S. dublin, S. montevideo and other serotypes. They all produce similar clinical disturbances.

(1) Abortions mainly in the latter half of pregnancy, although apparent barren ewes may reflect earlier foetal loss.

(2) Except for S. montevideo, ewes often ill at/or before the time of abortion: some scour, some die, some have an offensive smelling vaginal discharge for a week or more.

(3) Except for S. montevideo, lambs often ill at birth or subsequently, and developing a fatal septicaemia or pneumonia.

(4) Symptomless carriers in both ewes and lambs.

(5) A measure of immunity is apparently induced because it is uncommon for ewes to abort the next season.

S. abortus ovis is confined to the southern half of England, and particularly to Devon and Cornwall and even here it is now extremely rare. As it appears to affect sheep only, and has certain specific bacterial growth characteristics, specific vaccines may be necessary for its control.

Diagnosis

(1) Salmonella is usually easily found in membranes and foetus, providing faecal overgrowth is avoided (a possibility if only vaginal swabs are provided).

(2) Blood testing and rectal swabs provide very little worthwhile information for the practitioner.

Treatment and Control

Salmonella is frequently found to be sensitive to a range of antibiotics in vitro and theoretically treatment of individual sick animals is indicated and also carriers (which usually means the whole flock). In practice such treatment frequently fails to cure the sick or sterilise the carrier, and is expensive. Furazolidone (Neftin) has been used both in feed and by hand dosing, but without proven value. Consideration must be given in any salmonella outbreak to the risks of infecting man, and to producing drug-resistant organisms.

The principal aim in the control should be to:-

(1) Reduce the weight of infection by the isolation of aborting, scouring and sick sheep, and by high dose therapy of such animals if worth it.

e.g. 50kg (110lb) ewe - 250mg of furazolidone orally twice daily for 3 days. A 4% premix is available which can be mixed in the food to give the above doses for 7 days.

(2) Prevent as far as possible further infection of pregnant sheep, by keeping the aborting group (usually the nearest to term) separate from other groups yet to lamb.

(3) Upset the flock as little as possible and make sure food (hay or concentrate) is always available, particularly immediately following flock movement.

(4) Try and ensure as far as possible that most of the flock becomes immunised after lambing by spread from carrier ewes to non-pregnant ewes i.e. mix recovered aborters with ewes that have lambed. It may be valuable to produce a homologous vaccine.

(5) Avoid re-introduction of the disease from sources such as 'foreign' slurry, bought-in sheep and other grazing stock, and infected feed (including bird faeces). Attempts should be made to prevent birds, especially magpies, entering sheep houses and feeding from the sheep troughs.

FARMER'S FIRST AID CUPBOARD

Lubricant (e.g. Lubrel)
Arm-length poly gloves
Lambing cords and figure-of-8 telephone cable (or e.g. Hughes Lambing Snare -Arnolds)
Lamb stomach-tube and syringe (e.g. Arnolds)
Frozen colostrum (200ml lots)
Antibiotic injection (e.g. L.A. penicillin or L.A. oxytetracycline)
Antibiotic drench for lambs (e.g. Neobiotic P. pump)
Ophthalmic antibiotic (e.g. Aureomycin ophthalmic)
Calcium/Magnesium injection (400ml bottles of 20%)
Propylene glycol and glycerol (e.g. Ketol)
Tincture of iodine (e.g. in a spray)
Cotton Wool
Spirit
Needles 16 x 1", 18 x 1/2", 19 x 1"
Syringes 50, 10 and 5ml
Thermometer (to test lamb for hypothermia)

OBSTETRICS

General

It has been calculated that:-

(1) 70% of ewe deaths occur at or near lambing

(2) at least 75% of lamb deaths occur at or near lambing

(3) 90% of vets visits occur at or near lambing

These figures demonstrate that parturition is a high risk occupation and also that vets are 'dangerously' dependent upon this time for their contact with sheep farmers; but it is rewarding fun!

Uterine manipulation is relatively easy in most breeds, particularly for small hands, and limited embryotomy and caesarian section are simple.

There are particular problems associated with certain breeds and different problems in hill and lowland flocks, e.g.

Cluns	-	relatively small inactive lambs leading to post-parturient deaths
Kerry's	-	relatively large active lambs leading to parturient death
Texels	-	short neck and large lambs, especially in yearlings -do caesar early

The seasonal pattern of lambing presents problems, with sudden numbers and lack of practice, a large susceptible population and very intensive conditions.

Shepherd interference is the rule and the lamb(s) often dead when the Vet's assistance is required. The shepherd's job is usually quite different to the Vet's -the shepherd has a problem of numbers, timing (when to interfere), observation, cleanliness, and often very simple manipulations, principally the removal of 'tight' lambs in normal presentation and position, and to ensure the lambs are not smothered and/or mis-mothered.

Maternal factors

Ewes do not withstand prolonged vaginal and uterine interference; a sick ewe = a dead ewe!

Uterine, cervical and vaginal tearing is common following rough handling and inadequate lubrication.

Clostridial sepsis is common after vaginal interference if vaccination is inadequate and/or penicillin not injected.

Retention of the foetal membranes is rarely a problem.

Foetal factors

Often more than one lamb - a sorting out job - be patient.

Head and leg deflections common.

The head is the major cause of obstruction and there is rarely sufficient room in the pelvis for lamb's head and legs plus your hand, so use a snare over the head to permit traction.

Tight lambings often cause hepatic or intra-cranial haemorrhage leading to rapid death, or recumbent and hypothermic lambs which need urgent attention.

Equipment

Lubricant and small hands.

Antibiotics: Parenteral single-dose long acting penicillin or oxtetracycline plus antibiotic pessaries or powder.

Snares: 3 braided cords or binder twine with one labelled. Plastic lamber. Figure of eight telephone cable to act as 'snare', or Hughes Lambing Snare (Arnolds)

Sharp knife.

Examination

History: How long has the ewe been lambing?

Change in temperature	- 2 to 4 hours	Normal range
Cervical dilation	- 1/2 to 2 hours	
Delivery	- 1/2 to 2 hours	

Have any lambs been born or delivered?

External: Does the ewe look ill or exhausted?
Are there signs of shepherd interference? Note bleeding and bruising.
Presence of lamb/membranes at vulva.
Smell. Putrefaction implies bad risk, little value and embryotomy rather than caesar.
Udder. May suggest prematurity, mastitis and lack of colostrum.

Internal: Position the ewe for examination. Sometimes helpful to suspend the ewe.

Ewe: Pelvic size
Damage of vagina
Cervical dilation (Ringwomb)

Lamb: Dead or alive (not always easy to be certain - assume alive if not sure)
How many?
Presentation/position

Delivery

Aim to complete delivery within 15 minutes.
Positioning of ewe and lubrication very important.
Avoid straining and allow repulsion.

The Head:

The main bulk. May require snare or plastic lamber. It is common for a swollen head only to be presented outside the vulva, with both front legs "back". If the lamb is dead (no suck or blink) then decapitate; if alive, then suspend the ewe, lubricate and insert a small hand beside the head and neck and try to find a limb and extend it. It is often then possible in ewes (but not ewe-lambs) to withdraw the lamb with only one limb extended.

Repulsion:

Of limbs, with ewe suspended, often necessary, but be careful.

Posterior presentation:

Provided both hind-limbs and tail are present with-draw with gentle traction but beware of excess pressure on the lamb's rib cage causing hepatic rupture and fatal haemorrhage. If only one hind limb or a tail is visible, convert to proper posterior presentation and remove this posterior presented lamb first.

Snares:

It may sometimes be necessary before repulsion to secure the head and/or limbs with snares; trial and error.

Traction:

One handed - minimum force.

Embryotomy:

Decapitate or skin the limbs OUTSIDE vulva.

Caesarian Operation:

Indications (only when live lambs?):-

(1) Non dilation of cervix
(2) Foetal oversize
(3) Small pelvis
(4) Mal-presentation
(5) Torsion
(6) Deformities
(7) Pregnancy Toxaemia

After Delivery

(1) Re-examine for:

(i) presence of other lamb(s)

(ii) damage

(iii) membranes

(2) Antibiotic injection and intra-uterine pessary or powder

(3) Udder

(4) Attention to lamb(s)

Revival and survival - suspend, blow in nostrils, dry and warm, return to dam quickly for licking.

Colostrum - via dam or bank.

Navel - tincture of iodine spray or antibiotic aerosol.

N.B. Madel's rules for lambings = 1L + 4R'S = Lubrication, Recognition, Repulsion, Removal, Re-examination.

THE EWE: SOME GYNAECOLOGICAL PROBLEMS NEAR PARTURITION

Vaginal Prolapse (Eversion Vagina/Cervix)

(1) Usually lowland, old, fat, lazy ewes (several lambs).

(2) May be up to 20% incidence in a 'storm'.

(3) Usually a week or so before lambing (can be postlambing).

(4) Can be confused with lambing and urinating (straining).

(5) Uncertain cause (steep ground, over-fat, multiple lambs, bulk food, oestrogen from fungi in feed, coughing, short docking).

(6) Treatment: Clean the tissues and replace, Then retain by one of three common methods:-

(i) Truss - bindertwine, (see diagram), wire or plastic; or webbing and leather Devonian Ewe Truss for the more persistent case.

OR (ii) Wool tying (least useful).

OR (iii) Suture vulva (take big tuck, not too tight).

Binder-twine Ewe Truss

1. Tie round girth

2. Join together, knot at base of tail

3. Pass lateral to vulva, crossing below

4. Tie over abdominal spine

Also, remember to consider:-

(i) Reduce roughage
(ii) Moving to flatter field
(iii) Usually still to lamb, so keep an eye on her. Some cases have severely damaged vaginal mucosa and straining is persistent. Very efficient trussing is required, plus antibiotics and TLC.

Vaginal rupture/intestinal prolapse

(1) Intestines through vulva. Rupture usually near dorsum of cervix.
(2) Rapid death.
(3) Usually - late pregnancy, multiple lambs.
(4) Uncertain cause, but don't forget 'gut fill'!

Uterine Inertia

Usually secondary to Ca/preg.tox/dead lambs/mastitis. i.e. the ewe is ill.

Abdominal Rupture

Rupture of the linea alba. Preparturient. Old lowland ewes carrying twins or triplets. Poor condition. Reduce roughage. Be ready for caesar. Cull.

Uterine Prolapse

Usually post-parturient (immediate)
Not much shock
Remove membranes
Usually easy to replace
Clean up, suspend, antibiotic, truss and/or vulval sutures.

PERINATAL PROBLEMS

Many farmers accept that up to 15% of all lambs are either dead when born or die within the first few days of life (occasionally these losses rise up to 50% e.g. in abortion storms and seasonal swayback).
e.g. 1,000 lowland ewes anticipating 1,500 lambs may lose over 200 lambs in the perinatal period (0-10 days). These losses can be split into:-

(1) Pre partum stillbirths (abortions) 2%

Many of these will be due to uterine infections such as Toxoplasma, Salmonellae, Vibrio, Enzootic Abortion, Border Disease, plus Tick Borne Fever, Pregnancy Toxaemia and Swayback.

(2) Parturient stillbirths (dystocias) 5%

If the losses are due mainly to lambing problems then the quality and quantity of the work force needs looking at, and also the age, breeding and feeding, e.g. large singles, and ewe-lambs that are C.S.4 and crossed with Texel ram!

(3) Post partum mortality 8%

Often said to be born dead, but actually died soon after birth.

It is important to try and split lamb deaths into these 3 groups by fresh post mortem examination, in order to apply sensible control measures, e.g. if the deaths are mainly pre partum one has to look for causes of death in-utero, e.g. abortion; pregnancy toxaemia; if at birth then the shepherd's vigilance may be inadequate (e.g. not over 24 hrs.), if after birth, then check on warmth and colostrum.

Post Mortem Features (on 'Fresh Neonates')

(1) Decompostion -

indicates pre partum death

(2) Weight -

singles 4.5kg; twins 3.5kg (allow for breed variations)

Excessive weight suggests dystocia, underweight suggests pre partum problem (feed, infection)

(3) Umbilical vessels -

a blood clot indicates that death was partum or post partum.

(4) Navel cord -

a shrivelled cord indicates post partum death, square end indicates partum and tapered end indicates pre partum death.

(5) Brown fat -

changing from normal pink to brown gelatinous/absence, indicates post partum death (and starvation/hypothermia)

(6) Presence of food in stomach -

indicates post partum death

(7) Lung aeration -

indicates partum or post partum death

(8) Subcutaneous oedema of distal legs/tail and head -

indicates dystocia or hypothermia

(9) Hepatic rupture, thoracic, abdominal and meningeal haemorrhage (open up cranium and spinal canal) -

indicates dystocia

Post Partum Mortality

(1) This may still be an expression of PRE PARTUM disturbances

e.g.	soft lambs	-	Enzootic abortion
	hairy shakers	-	Border Disease
	ataxia	-	Daft Lamb, Swayback and Muscular Dystrophy

It is often an expression of poor pre partum management of the ewe e.g. improper feeding of the ewe in late pregnancy leading to:-

(i) Short gestations, giving small weak lambs with poor protective fleeces and little brown fat reserve. Such lambs become hypothermic and unable to rise. They require a bit of luck with the weather, and a good mothering shepherd, if they are to survive.

(ii) Less colostrum and milk, less sucking, and a vicious circle.

(iii) Poor mothering.

(iv) Pregnancy toxaemia.

(v) Copper and Selenium deficiencies - Swayback and Muscular Dystrophy.

(2) Sometimes the losses are due to a mixture of POST PARTUM infections introduced orally, or via the navel, or via early docking and castration wounds.

(i)	Enteric (usually first few days)	-	Lamb dysentery Watery mouth E. coli Salmonella Rotavirus Cryptosporidia (protozoan parasite)
(ii)	Navel (first week or so)	-	Staphs., Streps., E. coli, Fusiforms, Clostridia (chauvoei 'complex') C. pyogenes.
(iii)	Castration/ docking	-	Clostridia (including tetanus)
(iv)	Orf	-	see later

Such infections produce a variable range of CLINICAL SIGNS such as:-

Dull, sleepy, weak, recumbent
Salivation
Scour
Navel Ill
Joint Ill
Central Nervous signs (e.g. fits, paralysis, ataxia)
Ill thriving (e.g. liver and kidney damage)
Scabby lips (and ewes with sore teats, leading to severe mastitis and then hungry lambs)

However, farmers (and sometimes vets!) are inclined to place too much significance on pathogens and therefore on therapy, whereas the main problem more often stems from susceptible lambs (small, lack of colostrum, cold and inadequately immunised mothers) and/or dose of potential pathogens (dirty conditions, instruments and hands, inadequate observation and treatment of individual cases and no isolation).

As a general rule, offspring survive if they are average birth weight, kept warm and receive adequate colostrum, whereas those that are small and cold and fail to obtain adequate colostrum, die.

Treatment

(1) Hypothermia

If a very young lamb is weak and with a temperature of less than 100°F (38°C) it requires firstly drying (if wet), preferably with a dry towel, then warming and feeding (warm colostrum if less than 12 h., or warm milk if over 12 h.). If it is collapsed (not able to sit up on sternum) (but avoiding the watery mouth and scouring lambs), give it an intra-peritoneal injection (just behind and to one-side of the navel, with the lamb held up by its front legs) of 25-50ml of warm 20% glucose (1/2 normal commercial strength) and parenteral antibiotic e.g. 1/2ml L.A. Terramycin. When the lamb is conscious and able to suck, feed 150 - 200 ml at least 3 times a day via bottle or stomach tube until the lamb is able to suck the ewe vigorously. If the lamb is one of twins, remove the other from the ewe and return them together when ready.

Warming a lamb is best done with a domestic blower in a constructed 'box', and a battery operated heater is available commercially.

(2) Scouring

As for all species, it is vital to provide isolation, warmth and plenty of fluids e.g. Liquid Lectade (Beecham); antibiotics aiming at *E. coli* e.g. Clamoxyl Oral Doser (Beecham) are also indicated but of less certain value.

(3) Navel/Joint Ill

Vigorous systematic treatment - high dose twice daily with broad spectrum antibiotics; valuable lambs merit joint perfusion.

Control

By attention to the following:-

(1) Break in the lambing sequence (batch lambing).

(2) Change lambing pens/areas (even re-strawing of area is helpful), turn out.

(3) Provision of adequate shelter (even a zig-zag of straw bales in the field is helpful).

(4) Cleanliness - instruments, hands, bedding and disposal of foetal membranes/carcases.

(5) Spraying navels with tincture of iodine.

(6) Early detection and isolation of cases, e.g. abortions, scouring.

(7) Ensure adequate warmth and a good supply of colostrum obtained from ewes (give 1 ml oxytocin and wait 15 min) or cows.

(8) E. coli vaccination ('Coliovac', Hoechst), price approx. 20p/dose. It has a schedule similar to clostridial vaccination and can be given simultaneously.

(9) Cull ewes that are thin and have faulty teeth and/or udders; feed the others properly.

(10) Review future policy for vaccination, feeding and the provision of copper and selenium.

Investigation

It needs time and thought and should include:-

(1) History -

numbers involved (size of problem) problems in previous years. Purchases and related incidence of disease (is the problem confined to one group e.g. replacement ewe-lambs). Breeding and lambing dates. Weather conditions throughout pregnancy. Feeding of ewes, vaccinations and dosings (e.g. copper, fluke). Farmer's account of mortality, clinical signs/treatment.

(2) Farm observations -

cleanliness, drugs and instruments, shelter, and isolation and warmth, feed available including colostrum, (do your observations agree with the farm story?), Shepherding - quantity and quality.

(3) Clinical examinations -

ewes - age, condition (handle), udders, weak lambs - age, size, clinical signs, castration/docking.

(4) Laboratory -

plenty of fresh material (including foetal membranes). It may include paired blood samples from marked ewes for enzootic abortion, toxoplasma and BVD, plus blood biochemistry (copper and GSHPx).

The work is very rewarding (even if you don't get paid for it!).

Watery Mouth, Slavers, Rattle Belly

These are common terms used to describe a condition which affects very young lambs (up to 3 days old) characterised by the following:-

1. Lambs rapidly become dull and weak, look miserable and don't get up, and they become unwilling to suck.

2. They have cold, wet lips and muzzles from drooling saliva. (N.B. Most terminally ill lambs dying of other causes often drool saliva).

3. No faeces are visible (meconium) and the tail is dry.

4. The temperature is at first normal, but hypothermia follows.

5. The abdomen is relaxed at first and looks full, but later there is tympany and tenseness which causes 'rattling' when tapped or shaken (like a displaced abomasum case and may cause distressed breathing.

6. Without treatment, they usually die within the day, but some live long enough to develop scouring and even joint ill, just to complicate the picture!

7. The incidence can be frighteningly high (>20%), and is most common in twins or triplets out of ewes with condition score <3 and lambed indoors.

Cause

The current theory is that lambs which develop Watery Mouth have swallowed too little colostrum and too many E.coli in the first few hours of life, resulting in gut stasis etc. This theory seems to fit most of the observed events and leads to a rational approach to effective treatment and control.

Treatment

1. i/m antibiotic against E.coli.

2. Warm soapy-water enema (10-20 ml. via cut-down stomach tube).

3. Stomach tube three times that day, 100 ml. of glucose/electrolyte solution e.g. Liquid Lectade (Beecham's Animal Health) plus oral antibiotic (to match the i/m).

4. Leave the ewe and lamb(s) together, and keep warm.

Control

1. Reduce the dose of E. coli:-

   (i) keep things as clean as possible; bedding, pens, udders.
   (ii) Oral dosing of lambs at birth with a suitable antibiotic.

2. Increase the colostral protection:-

   (i) Feed the ewes well (avoid CS <3).
   (ii) Ensure good bonding (avoid moving ewe and lamb too soon after birth).
   (iii) Ensure the lamb(s) suck (avoid early castration and have good shepherding).

Some farms with a history of high incidence, now give all lambs shortly after birth 100-200 ml. of cow colostrum. Unfortunately, the current E.coli vaccine (Coliovac, Hoechst) appears not to provide protection, the K99 antigen does not appear to be involved.

MASTITIS

It is perhaps significant that the 1983 edition of these notes contained no section on mastitis; this was because we felt that there was little useful to say to the clinician on the subject because of both lack of concern and good data. This is now changing and thankfully the pattern of disease is becoming better understood and control measures more authentic, although still a long way behind our knowledge of mastitis in the dairy cow (see 'Notes for the Dairy Practitioner', Liverpool University Press 1985).

Two forms of the disease are apparent:-

(1) Healthy ewe with chronic, palpable mastitis = grade 2C

(2) Sick ewe with acute, often gangrenous mastitis = grade 3

Both forms are irreversible and the first requires culling and the second life-saving.

1. <u>Chronic Mastitis (2C)</u>

This is usually discovered before tupping time at a time when the ewes are selected for breeding or culling (e.g. thin, teeth, barren, known disease). One half of the udder, less commonly both, is lumpy and distorted with palpable chronic abscesses. The prevalence varies between 1% and 15%. If these are not spotted at this time they become apparent at lambing and probably account for many of the ewes which lamb-down with only one functional gland, and for all the problems of neo-natal disease that this invites. It is uncertain when infection occurs, but it is assumed that it is at weaning time, because long-acting antibiotic, introduced then, is known to reduce the disease incidence. This is very similar to the dairy cow story and the range of organisms appears to be much the same, with staphylococci the favourite.

<u>Control</u> is suggested in the following ways:-

(1) Dry Ewe Therapy (DET) at weaning - with, for example 1/2 tube of Streptopen DC or 1 tube of Orbenin DC (Beecham) per gland. This is a two person job and the teats need to be cleaned before infusion.

(2) Place ewes on poor keep and reduced water at weaning and keep well away from the sound and sight of their lambs, to reduce milk production.

(3) It seems a reasonable idea to teat-dip with a commercial cow teat dip at the time of infusion.

(4) Colour tag cases for culling.

2. <u>Acute Mastitis (2A or 3)</u>

This occurs most commonly soon after lambing, when the ewe is milking well and being sucked fiercely. It seems sometimes to follow a period of rough weather, particularly cold and wet, and shorn or crutched ewes are perhaps particularly vulnerable shortly after turn-out or if placed on lush clover pasture. The clinical picture is dramatic and typical; the ewe is obviously ill with a swollen udder (usually one half), she walks stiffly and the lambs appear hungry, the teat and a portion of the udder skin often becomes cyanotic and cold, and oedema extends along the mammary vein. Slow sloughing often occurs if the ewe survives the first day or so, with an extensive loss of abdominal skin, leaving tubes of mammary ducts and blood vessels exposed; healing takes many weeks and the ewe loses much condition.

Staphylococci and _Pasteurella haemolytica_ are again the principal pathogens, but _Escherichia coli_ and _Corynebacterium pyogenes_ are also on the list, as well as, surprisingly, _Clostridium perfringens_. This range of organisms is interesting and hopeful because there are commercial _E. coli_, clostridia and pasteurella vaccines designed for use in pregnant ewes, and staphs. and _C. pyogenes_ are tailor-made for DET (compare the dairy cow). In addtion, orf on ewes' teats after lambing is a significant hazard, and vaccination at around tupping time might help to solve this problem.

Treatment

Early vigorous systemic treatment is indicated i.e. high dose of an antibiotic which is likely to cover the range of bacteria (you have plenty of choice!) and repeat twice daily until the ewe shows improvement (a day or so). Strip the gland out at this time, and drainage is aided by removal of the teat if gangrenous. Intra-mammary tubes are an optional extra. Ensure the ewe is comfortable, preferably in isolation, and has tempting food and water nearby. Theoretically, i/v fluids are indicated on day 1 (1-2 litres).

Control

The incidence varies between seasons with changing management, but is of the order of 1-5%, and can clearly warrant money spent on its control, as for example:-

(1) DET and teat dip (consider a trial group if thought too expensive to do all the flock).

(2) Ensure proper clostridia vaccination, and consider _E. coli_, pasteurella and orf vaccination for this and other reasons.

(3) Clean dry bedding at lambing ('Squelch' test!).

(4) Avoid turning out in cold wet weather and/or ensure shelter.

(5) If ewes are too thin (CS <3) consider increasing the feed and ensure adequate milk for vigorous lambs.

(6) Attend to teat sores - pustules, orf and sucking wounds and this may mean antibiotic injections and even weaning the lamb(s).

(7) Colour tag cases for culling

C.N.S. DISEASES OF LAMBS

SOME QUESTIONS OF HISTORY

Ask the farmer to describe the clinical signs.

When did the signs first show? At birth, 1 week + etc.

How many cases have there been so far this season?

How many normal lambs are there?

How many have died?

Do any get better? - with or without treatment.

Are the affected lambs out of one particular batch of ewes?

Are they single lambs?

Has the farmer 'done anything' to them recently' - changed pasture, weaned, dock/castrate, inject/drench.

Has the farmer any cases to examine today?

Are there ticks?

Ask questions about the ewes (see Adult CNS diseases)

SWAYBACK (Enzootic Ataxia)

This is one of the most important of the many central nervous diseases of lambs, and is caused by faulty development and degeneration of nervous tissue in the brain and spinal cord of lambs whose mothers had low copper concentrations in the blood in the latter part of pregnancy.

These clinical notes assume a knowledge of the pathology and diagnostic clinical pathology of the condition, as well as the toxicology of copper.

Diagnosis

1. <u>History:</u> Suspicion arises from one or more of the following:-

   (i) Swayback has been diagnosed clinically on the farm before.

   (ii) Copper has not been given to the ewes during pregnancy.

   (iii) It has been a mild winter (which influences the quality and quantity of the grass and soil eaten and the quantity of concentrate consumed). Housing greatly reduces the risk.
   N.B. Ministry forecasts are published in the farming and veterinary press.

   (iv) The grazing is lush (e.g. improved by fertilizer), cf. hypo-magnesaemia. Improved/reclaimed pastures are particularly likely to give rise to swayback with single-sward grasses and high Mo and S levels.

2. <u>Clinical signs</u>

   (i) <u>Congenital</u>
   Lambs at or within a few days of birth, show varying degrees of ataxia and inability to stand; some will be able to suck if permitted and appear alert and sensible. Some may be born dead, or die within a few hours, i.e. swayback is a cause of perinatal losses. The signs are, at best, only suspicious and can easily be confused with the other CNS disturbances of young lambs.

   (ii) <u>Delayed</u>
   Lambs a few weeks old, which were normal at birth, developing varying degrees of hind-leg weakness, ataxia (Swayback) and paresis. The signs are more obvious when the lamb is chased about. It looks like a spinal rather than brain condition and the lambs can suck and graze normally; some will fatten because they learn how to walk again.

(iii) Acute Delayed

Rapidly developing CNS signs in lambs a few weeks old and dying within a few hours (cf. CCN). Uncommon.

3. Incidence

This will depend upon the control measures adopted and the weather, but it is common for a swayback-prone farm to have a few years with very few cases, followed by a calamitous 'storm' (50%); fortunately, the advent of housing and safe copper supplementation has reduced these incidents.

4. Pathology: Material for diagnosis:-

(i) Fresh brains and brain stems.

(ii) At least 10 heparinised blood samples from pregnant ewes that have not recently had supplementary copper.

Concentrations below 0.07 mg/100ml (or below 9 umol./litre) suggest a copper problem and supplementation is required.

(iii) Liver copper concentrations - below 235 umol./kg D.M.

(iv) Herbage analysis: Cu - below 5 ppm D.M.
Mo - above 1 ppm D.M.
S - above 0.20% D.M.

Treatment

Because swayback can be a progressive condition, it is worth treating those young lambs not too severely affected together with their contacts, with a single copper injection or drench (5-10mg) in an attempt to stop the condition getting worse. There is also evidence that low blood copper concentrations occur in growing lambs on improved pastures, and that copper, given as copper oxide needles in a capsule - Copporal - Beechams (lamb dose = 13p), will improve weight gains in these lambs, even in the absence of obvious clinical signs e.g. thin bones with fractures and poor quality fleece.

Control

1. During an outbreak:

All pregnant ewes should be dosed with copper, together with those lambs already born out of ewes not previously dosed, in order to prevent delayed swayback.

2. Next year:

Supply copper, preferably by dosing all pregnant ewes with copper (and remove other known sources of copper, e.g. copperised licks). If there is a problem of ill-thrift in lambs due to copper deficiency, then the lambs require dosing (it is difficult to increase milk copper concentrations by dosing the ewe).

There is a variety of preparations available; they vary in their cost and speed of absorption and formulation.

Injectables - given to ewes in early or mid-pregnancy

(i) 'Cujec' (Coopers)

Water soluble, rapidly absorbed and therefore potentially toxic; hence the low ewe dose of 12mg; but non-irritant and given s/c in multi-dose pack, therefore very convenient (15p/50kg ewe). Short acting?

(ii) 'Copavet' (C. Vet) and 'Coppa' (Rycovet)

Suspension, slowly absorbed and therefore safe, (ewe dose of 40-45mg); but irritant and better given i/m via a syringe, therefore less convenient (approx. 20p/dose).

Long acting. Can also be given to lambs to prevent delayed swayback at $^1/_4$ of ewe dose.

Oral preparations

(1) There are two oral preparations which are safe and long acting.

(a) Copporal - Beechams. 2g (lambs) and 4g (ewe) capsules of copper oxide needles. Dosing by plastic gun, of ewes round about tupping, and to lambs over 3 weeks, costing 21p for ewes and 13p for lambs.

(b) Cosecure - Coopers. A mixed mineral (copper, cobalt and selenium) as a soluble glass bolus given annually to adult sheep (over 6 months), preferably shortly before tupping, with balling a gun, costing approximately £1, or to lambs over 4 weeks cost approx. 60p.

(2) Short acting drenches are often used, but they require repeat dosing and risk toxicity. Copacobal (C.Vet) contains both copper and cobalt and to prevent swayback, it is given in mid-pregnancy and repeated one month later, costing approximately 6p per dose (10 ml of the diluted preparation). Copper sulphate (1.0g in 30 ml water) is still used prophylactically by farmers, but it is arguably a risky material to have about on sheep farms; copper heads the list of poisons in sheep (79 incidents in 1985).

To avoid TOXICITY (either within hours of injection or weeks after dosing and following progressive storage):-

(1) make sure copper is necessary
(2) reduce dose of injection in small sheep
(3) remove any other obvious source of copper, particularly if housed and fed concentrates
(4) treat ewes with care, and follow makers' recommendations.

There is evidence that individuals and breeds vary in susceptibility to copper deficiency (e.g. Scottish Blackface and Swaledales more prone) and a change in breed might be considered, or an alteration in the lambing date, or special precautions for selected breeds, or cull those ewes that produce swayback lambs. Note that high sulphate and molybdenum (Mo) levels in the diet are known to decrease copper 'availability' (hence the term "conditioned" copper deficiency; and hence also the use of ammonium molybdate and sodium sulphate in controlling chronic copper toxicity), and they are usually involved if copper deficiency is blamed for ill-thrift, osteoporosis, grey faces and loss of crimp in lambs (not forgetting lack of food and/or parasites!).

BORDER DISEASE (B.D.) AND HYPOMYELINOGENESIS CONGENITA (H.C.) 'HAIRY SHAKERS'

A disease of lambs characterised by:-

(1) Hairy birthcoat which may be excessively pigmented in normally smoothcoated breeds of sheep.

(2) A flock history of unaccountable poor growth and viability in some lambs, a proportion of which may show rhythmic clonic spasms (shakers) i.e. chorea; the shaking may be very fine and is absent when the lamb is asleep.

(3) The occurence in lambs under six months of age of myelin defects 'HYPOMYELINOGENESIS' without low copper concentrations in the blood and without neurone damage.

(4) A high incidence of abortions and barren ewes with no other obvious cause.

Originally, Border Disease was reported in pedigree Kerry and Clun breeds principally in the Welsh Border counties and, at first, it was difficult to be certain that a collection of distinct entities had not been called one disease, e.g. a mixture of genetic fleece disturbances plus swayback plus a runt bunch of lambs due to feed/parasites. Transmission and vaccination experiments now indicate a virus infection of ewes in early (less than 2 $^{1}/_{2}$ months) pregnancy, and the disease is known to be widespread. The virus is closely related to or identical with that which causes Bovine Viral Diarrhoea (BVD - Mucosal Disease).

Clinical Picture

(1) A number of lambs showing a typical long hairy coat, some of which are also ataxic with rhythmic tonic/clonic contractions of skeletal muscles. These signs are usually evident at birth, although in the naturally long-woolled breeds, the abnormal birth coat is not obvious.

(2) The incidence of B.D. and H.C. lambs is very variable but 'storms' are reported. It presumably depends on (i) the stage of pregnancy when infection is incurred; (ii) the number not previously exposed to infection and (iii) the time since the last exposure to infection.

(3) Many of these lambs survive for a time, but their growth rate is poor and they appear stunted with domed heads and lag behind the rest of the flock. If any survive for breeding, they are usually infertile.

(4) Gimmers produce the highest % of B.D. and H.C. lambs, and although further crops from "infected" dams may show signs of B.D., the signs are less obvious, suggesting progressive immunity.

(5) When the prevalence of B.D. is high, there is often also a high prevalence of infertility/abortion, or even of just weak lambs. This means that the disease may show as an abortion and infertility problem rather than as a disease of young lambs.

Material for Pathology

(1) A live lamb or formalinised brain and spinal cord. Even those lambs not shaking will often show histological signs of a sub-total failure of myelination of the brain and spinal cord which is distinguishable from swayback.

(2) Bloods for serology, but beware of lambs with negative titres and yet excreting infection, because of immune-tolerance.

Treatment

None, other than good management of affected lambs, which can help to mask the disease.

Prevention

Apply abortion control measures, and if possible, maintain a closed flock. Mix bought-in sheep (preferably as lambs) with the main flock as long as possible before tupping. Mix affected lambs and dams of affected lambs with non-pregnant sheep to promote immunity before next tupping, but do not breed from affected sheep because they are likely to be persistent excretors of the virus. Vaccination is a possibility, but there is no preparation available at present. Avoid mixed grazing of cattle and ewes in early pregnancy.

DAFT LAMB DISEASE - CONGENITAL CEREBELLAR CORTICAL ATROPHY

A non-febrile disease of the CNS of newborn lambs characterised by faulty cerebellar development and function. Probably hereditary and a particular problem in Border Leicesters. May also be an expression of a maternal infection during early pregnancy (cf. Border Disease).

Distinctive Features

(1) Usually seen at birth.

(2) Mentally stupid (daft).

(3) Head nodding and swaying and opisthotonus with the head arched over the back. Rarely still, aimless wandering.

(4) Histology - Degeneration of Purkinge Cells and Golgi Cells in Cerebellum -myelin normal and normal copper concentrations in the blood.

Incidence

Usually low, but could justify changing the ram, and not breeding again from the dams of affected lambs.

NUTRITIONAL MUSCULAR DYSTROPHY (NMD)

Vitamin E and Se Deficiency (VESD), White Muscle Disease (WMD), Stiff Lamb Disease

This is essentially a disease of young lambs which is becoming more common as a result of changing husbandry and feeding. Deficiency of either Vit. E or Se or both causes sudden muscle weakness and lameness (particularly hind-legs) and sometimes sudden death (another cause of 'found dead') in otherwise healthy lambs. The increasing tendency to feed only home-grown foods, to add preservatives to straw and cereals (alkali and propionic acid) and to artificially fertilize with high N and SO_4, promoting rapid grass growth, have all led to overt clinical disease now appearing in areas which, as judged by river-bed surveys, have been short of Se for a long time.

Fortunately, the Se responsive ill-thrift in lambs reported in other countries and the early foetal death producing barren ewes and the increased susceptibility to disease because of reduced phagocytic activity, have not been confirmed in the United Kingdom.

Clinical signs

The usual story is that thriving young lambs (50% occur 0-30 days and 25% 30-60 days) have just been turned out to fresh spring grass and within a few hours or days of running around, a few are found down and reluctant to get up; one or two may also be found dead. Those that are able to stand if forced to get up (you or your dog need to go into the field to find this out), take a few tottering steps and lie down again, looking alert and comfortable. Some may show distressed respirations due to cardiac failure, and also pneumonia is a secondary risk. This situation is, at first, easily confused with pulpy kidney, pasteurellosis, swayback, polyarthritis, spinal abscess and injury.

Diagnosis

(1) Suspicion arises if there has been previous evidence of Se deficiency on the farm and, in particular, if the flock is receiving little concentrate and no vitamin/mineral supplement, and also if a lot of roots and/or poor quality roughage is being fed.

(2) The absence of obvious signs of injury and arthritis; the sudden onset in several otherwise healthy looking lambs should point to NMD.

(3) Necropsy should eliminate pulpy kidney disease and pasteurellosis, but note that the muscle lesions are bilateral (and therefore no comparisons possible within the lamb - you need to be able to visualise what you are looking for!), and that if there is cardiac involvement, the lungs may look pneumonic.

(4) Serum (red tubes) from affected lambs - Creatine kinase (CK) over 1000 I.U./ml = muscle damage (but must be taken very early, as concentrations drop in a few days).

(5) Whole blood (green tubes) from 10 contacts - glutathione peroxidase (GSHPx) less than 20 units/ml (= 1 umol/litre or 30ng/ml Se) is suspiciously low.

(6) Inject with Vit. E and Se and see what happens! (often one is also forced to play safe and inject with antibiotics simultaneously).

Treatment

(1) Inject with Vit. E and Se e.g. 1ml 'Dystosel' (approx. 30p) and repeat next day if not much improved. Usually there is a good (diagnostic) response if caught early.

(2) Rest (bring in).

(3) i/v methylene blue (1% solution) - 1ml/4.5kg body wt.

Prevention

1. Ensure adequate levels of Se and Vit. E in the diet in late pregnancy:

   (i) Mineral/Vit. supplement containing 14mg of Se and 1500 I.U. of Vit. E per kg. and fed at rate of 15g/head/day.

   (ii) Inject 'Dystosel' once or twice in late pregnancy - 2-3ml costing 60-90p.

   (iii) Various drenches (and licks) e.g. Panacur SC, Pardevet. These tend to be an expensive and inaccurate way of dealing with the problem, and risk Se toxicity if mis-diagnosed.

2. Short-term control in lambs by injecting lambs at approx. 3 week intervals with e.g. Dystosel 0.5 - 1ml.

3. Long-term control by:

   (i) An annual injection (s/c in neck) to weaned lambs and pregnant ewes of a long-acting barium-complex of Se, at the rate of 1mg Se/kg e.g. Deposel (25p - £1).

   (ii) Dosing weaned lambs and ewes once every three years with long-acting pellets e.g. Permasel S (66p).

   (iii) "Cosecure" (Coopers) - see 'Swayback'.

   (iv) In drinking water as sachets e.g. Aquatrace, needing a piped water supply and preferably in troughs.

These long-term measures have not yet been properly evaluated in this country.

TICKS AND TICK-BORNE DISEASES

TICK CYCLE AND CONTROL

Twenty one species of ticks have been found in domesticated and wild animals and birds in Britain. The only one of importance in sheep is Ixodes ricinus, the so-called sheep tick, though it is not strictly host specific. It has a 3 host lifecycle in which less than 3 weeks of the 3 year cycle is spent on the host. Larvae and nymphs can develop successfully on birds and small wild animals but the adult needs a large animal host (sheep, cattle, deer). Since 98% of the lifecycle is spent off a host the success of the tick is very dependent on the external environment.

Geographical distribution

Found on rough hill grazing, in Scotland, the Pennines, Lake District, Wales, Devon and Cornwall and a few other small areas. These areas have an annual rainfall of over 100cm (40ins.). The vegetation is bracken, heather, rushes and coarse grasses and the number of ticks increase with the thickness of the vegetation matt.

Seasonal distribution

In most areas there is a tick rise in spring and another in autumn but the precise timing varies from year to year.

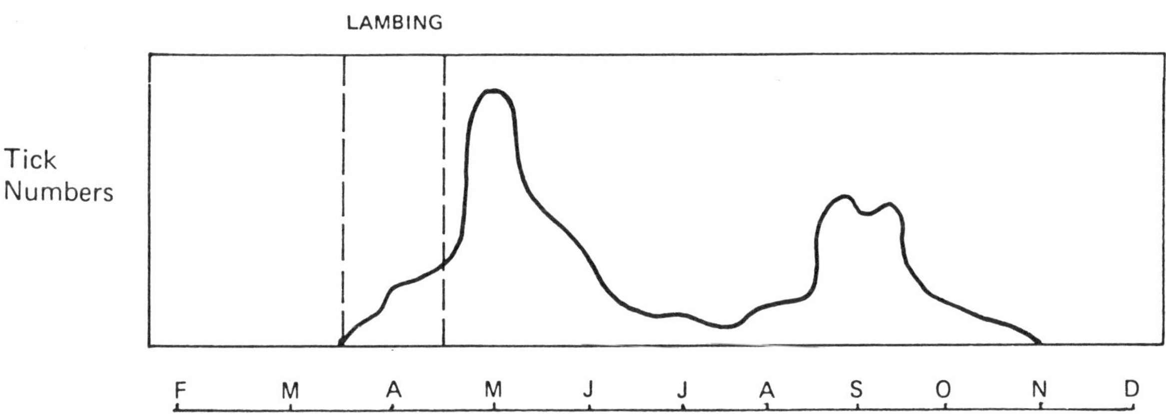

Small numbers of I.ricinus can be found on sheep from April to October apart from the peaks and in a few areas (Northumberland and south-east Scotland) there is no autumn rise. Spring rise and autumn rise ticks are different populations.

Flock distribution

The number of ticks found throughout the season on sheep increases with the age (or the size) of the sheep. In one count on a heavily infected farm, ewes had 400 adult females, hoggs 300 and lambs 50, together with about 5 times as many nymphs as adults, and 1,000 times more larvae than adults.

Sheep distribution

Ticks attach initially to sheep on the ears, nose, lips and feet and they migrate to become attached to their hairy parts with only very small numbers in the woolly areas. Adults migrate further than nymphs and larvae hardly migrate at all.

Importance

Ticks are reponsible for irritation and a certain amount of blood loss, particularly in young lambs (tick worry) and for the transmission of tick-borne fever, tick-bite pyaemia and louping-ill. The tick-transmitted diseases are more important than tick-worry because heavy infections are uncommon and the blood loss is over a period of some weeks, which allows for compensation.

Control

(1) Dipping

Ixodicides are mainly organo-phosphorus compounds e.g. Chlorfenvinphos, Dioxathion, Chlorpyrifos (e.g. Tick Dip Liquid - Cooper, Youngs 'Killtick') though Boots 'Amitraz' (Diamidide) is not. Chlorpyrifos has been used effectively as a spray on land to kill ticks in spring. The cost is about £12/ha.

Most tick dips are suitable for the control of blowfly, lice and keds and may also be approved by MAFF for sheep scab control. Most of these preparations give protection for 4 weeks in ewes though some claim to give partial protection for up to 8 weeks. Lambs are protected for only 1-2 weeks.

Ewes are dipped to reduce the pasture infestation in successive years and lambs are protected to try and reduce the number of ticks which they pick up. It is recommended that all ewes should be dipped either before lambing at the end of March or beginning of April or preferably after lambing before going to the fell. All lambs should be dipped or rubbed with salve before going to the fell. Lambs may be dipped from 4 days old, preferably by hand in a 45 gallon drum. Mismothering is prevented by:-

(a) allowing an hour after gathering and before dipping.

(b) drawing off batches of about 20 ewes with their lambs

(c) allowing lambs to drain and mother-up before putting a fresh batch through the dip.

They should be re-dipped in autumn.

'Pour-on' preparations of 2.5% cypermethrin (e.g. Ciba-Geigy 'Parasol' and Rycovet 'Ovipor') are applied to the sheep with a special gun in a line from the crown of the head to the top of the rump. The drug is not active systemically but spreads through the fleece. It is particularly useful in lambs (dose 5-10 ml. costing 10-20p) and gives protection for 3-4 weeks.

(2) Pasture Improvement

Many hill farms are draining and re-seeding the relatively low pastures bordering on the high hills in order to improve lambing percentage and live-weight gains in the lambs. This has a dramatic effect on tick populations which rapidly disappear. This will lead to a waning of immunity to tick-borne disease (especially louping-ill) and disease may occur when the sheep are later turned on to the hill grazings.

Monitoring of tick areas

Pastures and hills can be dragged during the time of the tick rise to detect ticks and estimate their numbers. A woolly blanket (white or pale-coloured) is dragged (see diagram on next page) adapting its shape to the vegetation and the ticks can readily be seen on the under surface.

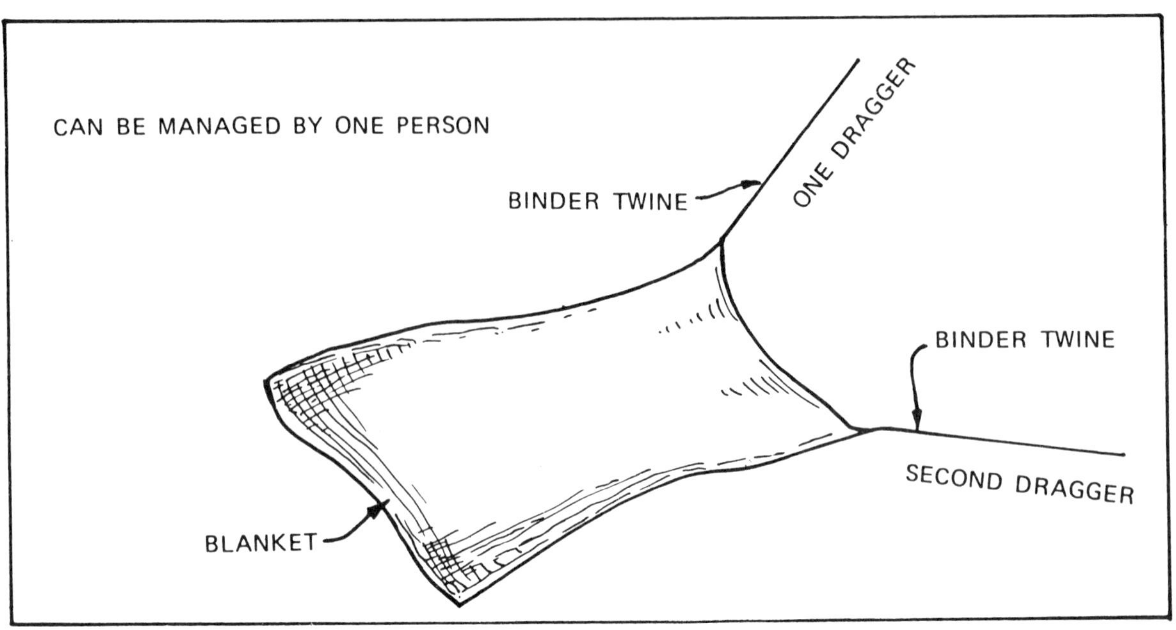

TICK-BORNE FEVER (TBF)

This disease of sheep is caused by a rickettsia, Rickettsia ovina phagocytophila (also called Cytoecetes phagocytophila or Erlichia phagocytophila). A similar organism of close antigenic identity is found in cattle. The organism is transmitted by the nymph and adult of I. ricinus and virtually every tick is infected.

The rickettsia invades the circulating neutrophils and monocytes and results in their destruction, thus causing a neutropenia. It is this feature, combined with a pyrexia, which gives rise to clinical signs and predisposes the sheep to staphylococcal pyaemia and increases the severity of louping-ill.

Clinical Picture

Since all ticks are infected, lambs in a tick area become infected very early in life. The clinical signs are mild and are usually not noticed - the lambs may be listless, fail to suck and have a temperature rise (up to 41°C) for a few days. There is some evidence that some strains are more pathogenic than others. Organisms can be found by Giemsa stain of a thin blood film. Immunity then develops, is probably short-lived but is reinforced throughout life producing 'acclimatised' sheep. There are no antibodies to TBF in the colostrum. More severe signs are seen in adult sheep brought from a tick free area. The high temperature which follows exposure to ticks may result in abortion in pregnant ewes (unusual because the time of year is usually wrong) and temporary infertility (for about 3 months) in rams. This is of considerable importance since rams are usually purchased at September/October sales and are infertile just at the time that they are turned out with the ewes.

Control

The organism is susceptible to sulphonamides and broad spectrum antibiotics. It is more important to obtain satisfactory immunity in lambs and to produce acclimatised sheep. Do not move non-immune ewes into a tick area during pregnancy. Work at the Leeds VI Centre has shown that lambs can be immunized against TBF by injecting blood from carrier ewes. Citrated blood is taken from 3 ewes which have been on the fell, which, it is assumed, will be carrying small numbers of organisms, since the immunity is a premunity and ewes remain as carriers for 2 years after experimental infection. The blood can be kept in a refrigerator at $4^{o}C$ for up to 1 week, and 1ml is injected s/c into the lambs (as young as a few hours old). They are then kept off the tick areas for 3 weeks which separates the neutropenia caused by TBF from the infection with staphylococci responsible for tick-bite pyaemia. Although field impressions over several years suggest that this routine reduced pyaemia/cripples, a recent critical controlled trial in two areas in the Pennines has failed to produce conclusive differences in pyaemia/cripples between those lambs which received the injection and those which did not. It was also clear that measures directed against the tick (dipping, heather burning) were much more effective than the injection of blood containing the rickettsia of TBF.

TICK-BITE PYAEMIA

A generalised staphylococcal infection of lambs of 2 to 16 weeks old associated with tick bites, and usually predisposed by tick-borne fever. Neutropenic lambs have been shown to be highly susceptible to Staphylococcal septicaemia. The disease is either a septicaemia resulting in death or the bacteria localised in many organs, producing abscesses in joints, CNS, liver, lungs, etc. Lambs are rarely infected before two weeks of age, probably because the neutropenia from tick-borne fever is not evident before this time, and also because adequate titres of staphylococcal antibodies derived from the colostrum wane at about two weeks.

Clinical signs

These vary with the site of the abscesses and include lameness with painful enlarged joints, paralysis of the hind legs due to abscess in the spinal cord, signs of meningitis or unthriftiness due to abscesses in liver, lungs, kidneys etc.

The septicaemic form shows in sudden deaths. The disease occurs wherever _I. ricinus_ is found and is of great economic importance. It has been shown that half the lamb losses on a tick farm can be due to pyaemia and that virtually all the lameness in lambs is associated with tick-bite pyaemia.

Treatment

Antibiotics (penicillin) are effective against the organism and if lambs are treated early, the clinical signs may regress and the lambs recover. Lambs with spinal abscesses or with multiple abscesses in liver and lungs are unlikely to recover despite treatment and it has been suggested that penicillin only hastens the recovery of lambs which were likely to recover spontaneously.

Control

(1) Tick control and control of tick-borne fever.

(2) Long-acting penicillin or tetracycline have been employed prophylactically just prior to exposure to ticks, when the lambs are sent with the ewes to the fell.

(3) Staphylococcal toxoid. Two doses before exposure might be useful, but lambs do not respond well before six weeks of age. By vaccinating the ewe during pregnancy, it has been possible to induce passive immunity in the lamb for only two weeks. Thus this method is of little practical value.

LOUPING-ILL ('Trembling')

This is a virus disease which affects sheep and many other hosts, including cattle, deer, man and red grouse. Typically, there is an initial viraemia about 2 to 6 days after the tick bite, during which there is a febrile reaction (42°C). The majority of acclimatised sheep show few clinical signs and even susceptible sheep will usually recover rapidly. In some sheep, however, the virus invades the CNS and severe clinical signs and death follow. There is evidence that a primary TBF infection at the same time increases the likelihood of CNS invasion. The virus is transmitted stage to stage by I. ricinus and only a small proportion of ticks is infected. Some tick areas have no louping-ill virus present.

Clinical Signs

Disease is seen in all ages of sheep from as early as 2 days old (? prenatal infection) to aged ewes. Lambs are dependent on colostral immunity for protection, which is strongest when immune ewes are exposed to re-infection a few weeks before lambing. Changes in grazing practices may prevent this natural boost to the immunity of the ewes and many lambs may then become clinically affected. Lambs may be found dead but the typical signs are those associated with CNS involvement and are variable in nature. They include resting the head on the ground, head-pressing, an abnormal jerky gait ('leaping'), staggering, circling and paralysis. Fine tremors of the facial muscles and ears, nystagmus and twitching or nibbling of the lips may be seen. The disease progresses within 1 to 4 days to coma and death, often assisted by predators.

Usually only a small proportion of the flock is affected but outbreaks may be explosive with a high mortality. Animals which have recovered from the viraemia have a solid long-life immunity.

Diagnosis in a severe outbreak is easily made on clinical grounds but CCN, listeriosis, swayback, gid or CNS abscesses must be considered. There is a typical histology of the brain and virus can be recovered at the Veterinary Investigation Centre by mouse inoculation. There is a serological test for antibody.

Treatment and control

Treatment is not effective but careful nursing indoors may save some cases.

The aim of control is to ensure that ewes are immune and that lambs are protected passively. There is a vaccine available (Louping-ill vaccine, (BP Vet, Coopers Animal Health) which costs £1.00/dose. The recommended procedure for ewes is that 1 ml should be injected subcutaneously a month before lambing with a boost every other year before lambing. It may be necessary to protect ewe-lamb replacements before the autumn tick rise.

A summary of possible control methods for ticks and the diseases they transmit is:-

(1) Dip ewes around lambing and on 2 occasions later in the summer to reduce tick populations.

(2) Dip, salve or 'Pour-on' lambs before returning to tick areas.

(3) L/A penicillin or tetracycline to lambs before returning to tick areas.

(4) Special care over introduction of new rams and ewes.

(5) Vaccinate ewes against Louping-ill every other year and lambs before the autumn rise.

(6) Heather burning of areas on the fells on a 7 year rotation.

THIN, SCOURING LAMBS

COCCIDIOSIS

About a dozen species of Eimeria have been described from sheep and their pathogenicity differs considerably. Many of them produce few if any clinical signs whereas a few species (E. crandallis and E. ovinoidalis) and mixtures of species, inhabiting different areas of the intestine, may cause severe disease.

History

The majority of outbreaks, which involve a high proportion of the lambs, occur in animals of 4 to 8 weeks old which have been housed since birth. Coccidiosis also occurs in lambs of about 6 weeks of age which have been turned out about 2 weeks previously and, less commonly, in lambs which have never been housed.

Epidemiology

The ewes act as a source of oocysts and sporulation occurs in 2 or 3 days in the bedding. Lambs become infected almost immediately after birth and they start passing oocysts in large numbers by 21 days old. It is these oocysts, after sporulation, which produce the clinical disease some 2 to 3 weeks later. After 6 weeks of age the lambs become increasingly immune and the number of oocysts in the faeces falls markedly.

Clinical Signs

The first sign is that the farmer may recognise that the lambs have lost 'bloom' and show an 'open' fleece, Some animals may have slight diarrhoea with soiling of the wool below the anus. These signs extend to a considerable number of the flock and weight gains are reduced. Severe untreated cases proceed to extensive scouring, dehydration and death.

Diagnosis

This is based on clinical signs and history.

Oocyst counts should be carried out on a number of faecal samples and will usually be over 100,000 per gram. It is possible, however, to see clinical coccidiosis associated with the developing sexual stages, before peak oocyst production occurs. Counts of up to 1 million oocysts per g. can be seen without clinical signs and care should be taken to consider other possible causes of scouring before assuming that coccidiosis is responsible. In lambs out at grass Nematodirus battus infection should be considered. It is useful to sporulate and identify the species of Eimeria but this is a specialist job.

Treatment

Sulphonamides

(a) Orally with 2.5ml/5kg body wt. of $33^1/_3$% solution of Sulphadimidine daily for 4 days (Coopers 'Sulphamezathine').

(b) Parenterally with 1ml/10kg body wt. of sulphamethoxpyridazine (Parke-Davis 'Midicel Parenteral') for one or two days.

(c) Amprolium/ethopabate. Orally with 1ml/10kg body wt. (Merck Sharpe & Dohme 'Amprol-Plus' Solution) daily for 4 days.

Control

Any measures which reduces oocyst numbers and defers the peak of oocyst production in the lambs until after 6 weeks old will be beneficial. Slatted floors, addition of dry litter and removal of wet bedding are helpful. On farms on which coccidiosis occurs frequently, strategic treatment with the drugs mentioned before may be given to lambs at 3 and 6 weeks old.

Monensin sodium (Elanco 'Romensin Premix' contains 10% monensin) at 15 ppm incorporated into the lambs creep feed from day old, markedly reduces oocyst numbers but is slightly unpalatable and this may result in slight reductions in weight gains. Monensin can also be given at 10 to 15 ppm in the diet of ewes for about one month prior to lambing and for 2 months after lambing. Oocyst counts are then very low and improved weight gains are seen in the lambs. This is only of practical value if the ewes are housed and are fed from the time of housing as oocysts will remain viable on pastures for up to a year. (It is not licensed for coccidiosis control in lambs in Britain but is available as a growth promotant in beef cattle.)

PARASITIC-GASTRO-ENTERITIS (PGE)

Although mixed infections with abomasal and small intestinal worms are usual, there is a peak of infection caused by different genera at different times of the year.

PGE is the important cause of thin, scouring lambs during their first grazing season but it may also cause disease in young adult sheep in the winter months.

The pattern of disease is usually as follows:-

May - June	Nematodirus battus
July - September	Type 1 ostertagiasis, usually together with Cooperia
October - March	Trichostrongylus
January - April	Type II ostertagiasis

N. filicollis can contribute from May to January and Haemonchus contortus may produce explosive outbreaks of acute anaemia and deaths during July and August.

Ostertagiasis

TYPE I: This occurs most frequently in lambs which are intensively grazed. Eggs are passed in the spring by the ewe (peri-parturient rise) or larvae overwinter and infection is then built up by the lambs. Weather conditions cause a 'telescoping' of larval development so that pasture infections become dangerous around July.

Clinical signs

Flock outbreaks occur with a profuse diarrhoea resulting in a soiling of the fleece around the tail. The wool lacks lustre and weight gain is reduced followed by loss of weight. Untreated cases lead to dehydration, weakness, recumbency and death.

Diagnosis

1. History and clinical signs.
2. Egg counts: usually over 1000 epg and often much higher.
3. Worm counts: usually at least 10,000 worms.

TYPE II: Hoggs and young ewes (i.e. animals around one year old) in winter months may show unthriftiness with intermittent diarrhoea. The animals may be housed or at grass and the condition closely resembles that seen in cattle and is caused by the sudden emergence of larvae from nodules after having been inhibited in development for several months.

Diagnosis

1. History and clinical signs.
2. Egg counts: variable but often over 1000 epg.
3. Worm counts: over 10,000 adults and many L4 inhibited in nodules in abomasum.

Treatment - see later

Trichostrongylosis

This often occurs with Type II ostertagiasis as the requirements for the development of the infective larvae on the pasture are similar but it also occurs as almost a pure infection somewhat earlier in the winter (November -December).

Affected animals (hoggs) have a dark-coloured foul-smelling diarrhoea ('black scour') and rapidly lose condition. Dehydration often leads to collapse and death.

Diagnosis

1. History and clinical signs.
2. Egg counts are often very high (over 5000 epg).
3. Worm counts: The intestinal species (*T. colubriformis* and *T. vitrinus*) are most important but *T. axei* in the abomasum may contribute. The total number of worms is usually over 30,000.

Treatment - see later

Control of Parasitic gastro-enteritis, excluding *Nematodirus battus* - see graphs

Since lambs are born at different times of the year depending on the system of husbandry it is necessary to relate control measures to the underlying epidemiological facts, which are summarised as follows:-

1. The sources of pasture contamination are two-fold -

   (i) The periparturient rise in ewe faecal eggs.
   (ii) Overwintered infective larvae.

2. A light pasture infection is built up into sufficient infective larvae to cause disease by passage through the lambs themselves.

The principles of control are therefore:-

(1) Anthelmintic treatment of ewes, mainly to prevent the periparturient rise in faecal egg counts but also for the benefit of the ewes themselves and a possible increase in milk yield. There is evidence that treatment of ewes at housing in January increased lambing percentage in March lambing flocks by an unknown mechanism. The number and timing of treatments after lambing will depend on factors such as when the ewes are housed and whether they are to be placed on 'clean' pasture i.e. pasture which has not had sheep on in the past year. One treatment around lambing time (or earlier if housed) is sufficient if clean pasture is available but if the ewes and lambs are on contaminated pasture the ewes should be treated 3 and 6 weeks after turn-out. By this time their immune response will be becoming re-established and their faecal egg count should be low.

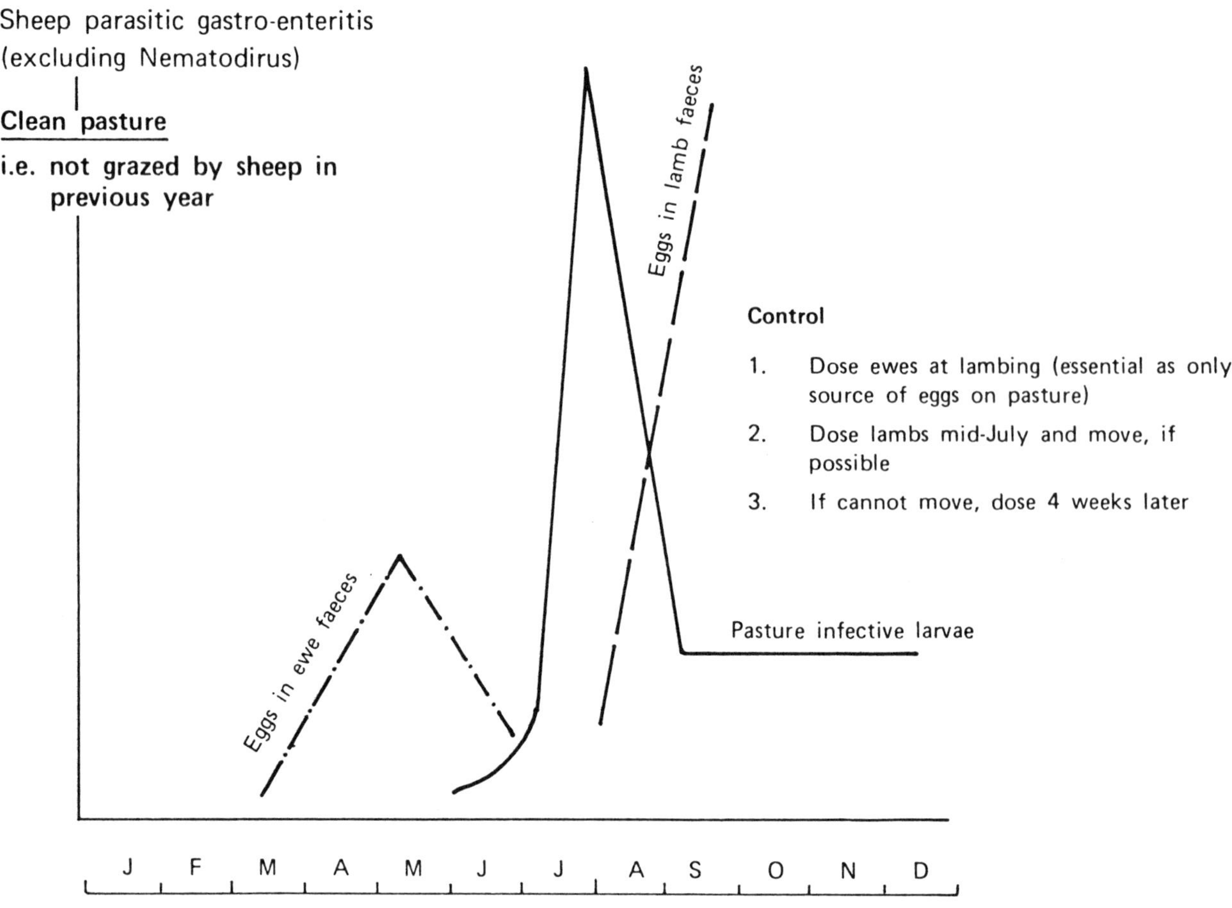
Sheep parasitic gastro-enteritis
(excluding Nematodirus)
Clean pasture
i.e. not grazed by sheep in previous year
Eggs in lamb faeces
Control
1. Dose ewes at lambing (essential as only source of eggs on pasture)
2. Dose lambs mid-July and move, if possible
3. If cannot move, dose 4 weeks later
Eggs in ewe faeces
Pasture infective larvae
J F M A M J J A S O N D

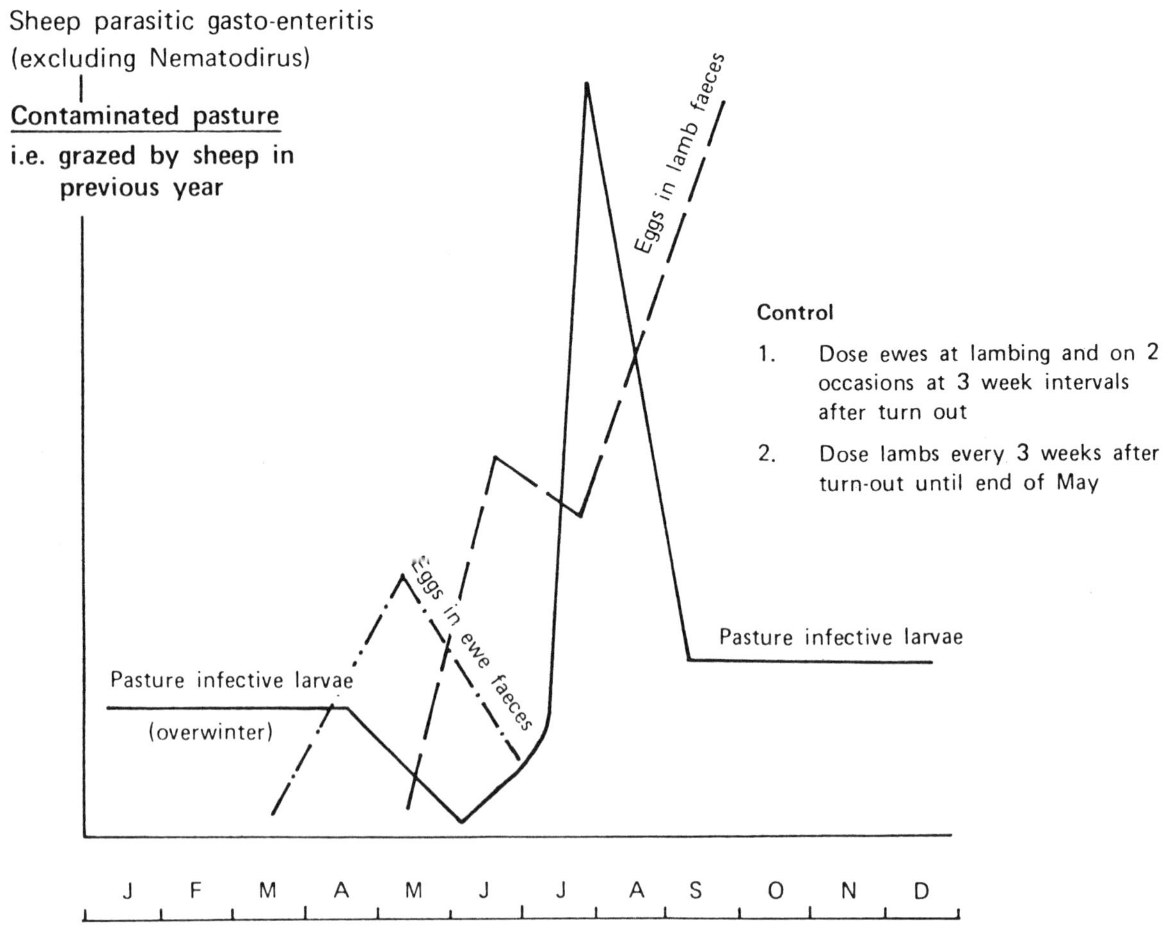
Sheep parasitic gasto-enteritis
(excluding Nematodirus)
Contaminated pasture
i.e. grazed by sheep in previous year
Eggs in lamb faeces
Control
1. Dose ewes at lambing and on 2 occasions at 3 week intervals after turn out
2. Dose lambs every 3 weeks after turn-out until end of May
Eggs in ewe faeces
Pasture infective larvae
Pasture infective larvae
(overwinter)
J F M A M J J A S O N D

(2) Anthelmintic treatment of lambs:

(i) If ewes and lambs are placed on clean grazing there should be no need to dose the lambs before they go for slaughter but a dose and move to clean pasture (hay or silage aftermath) in July might be advisable if lambs have not been slaughtered by this time.

(ii) If ewes and lambs are placed on contaminated pasture, lambs should be dosed every 3 weeks from 6 weeks old, since lambs do not ingest a significant quantity of grass until 3-4 weeks old, until the number of overwintered infective larvae has fallen to very few (end of May). The lambs should make good weight gains and the dosing should markedly reduce the pasture contamination later in the year (and, incidentally, the next year). This routine will also control Nematodirus infection. A 'pulse-release' or 'continuous release' device, given to ewes and lambs at turn-out would be an ideal method.

Nematodiriasis

Two species of Nematodirus - N. battus and N. filicollis are responsible for the disease which typically affects lambs 6-10 weeks old but can occur in older lambs. The life cycle of the two species differs and the epidemiology of the disease they cause is also different.

Typically, eggs of N. battus which are passed in April and May develop very slowly so that by the end of the summer, infective larvae will be present within the eggs. Hatching does not usually take place until the eggs have been subjected to freezing followed by a temperature of $10^{o}C$ so that mass hatching occurs in March to May, depending on weather conditions. The MAFF issue a nematodiriasis forecast each year based on soil temperatures. In essence, if warm conditions occur early (e.g. February/March) the eggs will hatch and the larvae will die off before lambs are consuming much grass but if the rise in temperature above $10^{o}C$ is late (May) severe disease will occur as lambs will ingest large numbers of hatched larvae. The forecast must obviously be interpreted in the light of lambing dates on a particular farm.

With N. filicollis, however, eggs passed in April will have hatched by June and large numbers of infective larvae will be on the pasture from September to May.

This means that N. battus (essentially found in the north of Britain) typically has a distinct seasonal incidence in May/June whereas N. filicollis can occur over a longer period. A considerable age resistance occurs with both species so disease usually occurs when young lambs and high pasture larval counts coincide. Although this pattern occurs in the majority of outbreaks of nematodiriasis, recent studies have indicated that both N. battus and N. filicollis may occasionally cause disease in autumn, perhaps by autumn-autumn transmission. Although all ages of sheep can harbour low numbers of worms and thus perpetuate the parasite, infections which will give rise to clinical disease are only produced by infected lambs.

Clinical signs

These are characterised by the sudden onset of profuse scouring, loss of bloom, dehydration and unthriftiness. A high proportion of the lambs are affected and losses may be as high as 30% (especially with N. battus). The disease continues for about 3 weeks and survivors are solidly immune.

Diagnosis

1. History and clinical signs.

2. Egg counts: This is not reliable at the start of an outbreak but frequently egg counts are over 1000 epg.

3. Worm counts: Mature and immature worms are found in the upper small intestine and usually number over 50,000.

Control

N. battus

1. Ensure that the current year's lambs do not graze on pastures occupied by lambs the previous year.

(i) Preferably by the use of new leys, OR

(ii) By grazing lambs on pastures on which only older sheep grazed the previous year.

2. If it is not possible to arrange grazing management so that clean land is available each spring for the lambs, strategic anthelmintic treatment must be used. This should be used in conjunction with the forecast but generally this involves a dose to the lambs late in April and two further doses at three-weekly intervals. The dosing scheme outlined earlier for PGE will automatically cover this period and thus control nematodiriasis.

3. If autumn nematodiriasis become a problem, prophylactic dosing of lambs in August/September may be necessary.

N. filicollis

The principles are similar to those described for PGE due to Ostertagia and Trichostrongylus (see later).

Treatment - See later. An increased dose of some drugs has to be given.

CLEAN GRAZING

Work in a number of centres in Britain in the past few years has shown that greatly increased grass production and, therefore, intensity of pasture stocking can be produced by nitrogen application on three occasions in March, May and June but this leads to severe parasitic disease unless the grazing is 'clean'. There are several systems which can be planned in a farm programme to increase the amount of clean grazing, either by annual alternation of cattle and sheep (preferably plus conservation and/or arable) or by the strategic use of anthelmintics to reduce the egg production of sheep grazing the pasture and thus the subsequent build-up of infective larvae. Clean grazing is not necessarily worm-free but infective larvae are few. Some of the ways to obtain and maintain clean grazing are indicated in the table. (See M.A.F.F. booklets 2154 'Grazing plans for the control of stomach and intestinal worms in sheep and in cattle' and 2324 'Clean grazing systems for sheep' for details.)

	BEFORE JULY 1	AFTER JULY 1
CLEAN	1. New grass after an arable crop. If grazed by ewes in previous autumn must have been dosed.	1. Aftermath not grazed by sheep earlier in year.
	2. Grazed by cattle in previous year.	2. Grazed by cattle only in the first half of season.
	3. Grassland used only for conservation in previous year (ewes grazing in autumn must have been dosed).	
SAFE (usually)	1. After May 1, grazed by yearlings in the previous year or by dosed lambs in the second half of the previous grazing season.	1. Clean pasture at the beginning of season grazed by dosed ewes and their lambs.
	2. Grazed by lambs before mid-March or after mid-September in previous year.	2. Safe pasture at the beginning of the season grazed by ewes and lambs, both of which have been dosed at 3 week intervals from turn-out to the end of May (or continuously).
DIRTY POTENTIALLY DANGEROUS	1. Safe pasture grazed by lambs in the first half of the previous grazing season, even if dosed on one occasion in May.	1. Safe pasture at the beginning of the season grazed by lambs in the first half of the current season.
	2. Safe pasture grazed by undosed lambs in the second half of the previous grazing season.	2. Safe or clean pasture at the beginning or the season grazed by undosed ewes in the first half of the current season.

ANTHELMINTICS FOR SHEEP

Chemical Name	Trade Name	Company	Dose mg/kg	Cost (p)* 25 kg lamb (approx.)	Formulation	Activity: G.I. NEMATODES	Activity: INHIBITED L4	Activity: LUNG	Activity: TAPE	Activity: FLUKE	Withdrawal periods days: MEAT	Withdrawal periods days: MILK
BENZIMIDAZOLES												
Albendazole	Valbazen 2.5%	S, K, F	5 (7.5 fluke)	8	Suspension (+Co/Se)	+	+	+	+	+ (Adults)	10	NU
Febantel	Bayverm 2.5% Amatron 2.5%	Bayer Coopers	5	7	Suspension, pellets	+	+	+	+	-	8	2
Fenbendazole	Panacur	Hoechst	5	8	Suspension (+Co/Se) powder, pellets, feedblock	+	+	+	+	-	14	3
Mebendazole	Ovitelmin	Crown	15	9	Suspension	+	-	+	+	-	7	NU
Oxfendazole	Synanthic	Syntex	5	9	Suspension, intra-ruminal injection	+	+	+	+	-	14	5
	Systamex	Wellcome	5	9	Suspension							
Oxibendazole	Widespec Anthelworm	Rycovet Young	10	5	Suspension (+ Co)	+	-	-	-	-	4	3
Parbendazole	Helmatac	S, K, F	20	8	Powder, pellets	+	-	-	-	-	6	NU
	Topclipworm	Ciba-Geigy			Suspension, pellets							
	Triban	Crown			Suspension							
Thiabendazole	Thibenzole	M, S, D	44	7	Suspension, paste (+Co), powder, pellets	+	-	+	-	-	0	0
Thiophanate (Probenzimid-azole)	Days' drench Nemafax	Day & Co. M & B	60	7	Suspension, pellets	+	-	-	-	-	7	3
OTHER COMPOUNDS												
Invermectin	Ivomec	M, S, D	0.2	10	Solution	+	+	+	+	-	14	NU
Levamisole	Many preparations e.g. Cevasol (Ceva), Cyverm, (Cyanamid), Duphamisole (Duphar), Levacide (Norbrook), Levadin (Univet), Bionem, Nemicide, Nilverm (Coopers), Ripercol (Crown)		7.2	6 - 10 depending on manufacture	Solution, granules	+	+	+	-	-	3	1
Diamphenethide	Coriban	Coopers	100	14	Suspension	-	-	-	-	+	7	NU
Nitroxynil	Trodax	M & B	10	4	Solution (20%)	-	-	-	-	+	30	NU
Oxyclozanide	Flukol Zanil	Young Coopers	15	5	Suspension, granules	-	-	-	-	+ (Adults)	14	0
Rafoxanide	Flukanide New Flucol	M, S, D Young	7.5	4	Suspension, injection	-	-	-	-	+	28	NU
Triclabendazole ‡	Fasinex	Ciba-Geigy	10	5	Suspension	-	-	-	-	+	28	NU

There are also a number of preparations which are a combination of drugs active against nematodes and fluke, e.g. duospec (Rycovet), Nemtrem (Young) - oxibendazole/oxyclozanide

Flukombin (Bayer), Vermadex (M & B) - thiophanate/brotianide
Nilzan (ICI) - levamisole/oxyclozanide
Ranizole (M, S, D) - thiabendazole/rafoxanide

* This price is given for comparative purposes. All these products are PML and the price is that given in the Index of Veterinary Specialities, October 1986.

‡ Although this drug is a benzimidazole it is placed here since it is active only against fluke.

NU not to be used for sheep producing milk for human consumption.

COBALT DEFICIENCY

Cobalt deficiency (the cause of 'pine' or 'ill-thrift') occurs in sheep, especially but not exclusively in lambs, and cattle in certain well-known areas. The incidence is highest in areas where soils are derived from acid igneous rocks and where there are coarse, sandy soils. In Scotland, cobalt deficiency is widely distributed but in England the localities most likely to be deficient are in the limestone areas of the Pennines, the Old Red Sandstone areas of Hereford, Shropshire and Worcester, Dartmoor and the Greensands at the edge of the chalk in south east England. As well as these areas where cobalt deficiency is endemic, other areas have become cobalt deficient as a result of farming practices, such as liming and reseeding which have improved pastures and in so doing, lowered the available cobalt. A constant intake of cobalt is needed to allow rumen organisms to produce vitamin B12 and the disease caused by cobalt deficiency is thus an induced vitamin B12 deficiency.

Clinical Signs

Characterised by loss of appetite, reduced weight gains proceeding to weight loss and extreme emaciation. Lambs have a dry coat and a tight skin. In the terminal stages, there is a severe anaemia and lachrymation. Subclinically, a marginal deficiency of cobalt may be of considerable economic importance since typical signs may not be present but weight gains may be reduced. Parasitic gastro-enteritis is often also present and copper deficiency may be a complicating factor causing reduced growth rate in hill areas where pasture improvement has been carried out. Farmers tend to suspect deficiency states with little evidence so care should be taken over diagnosis.

Diagnosis

1. History of farm and geographical area.

2. Clinical findings.

3. Response to treatment. This is often very marked after cobalt administration but it should be noted that lambs with reduced appetites due to other causes may also show a marked response to dosing with cobalt.

4. Laboratory confirmation:

In the animal (sample at least 10 animals)

(i) Serum vitamin B12 concentration

Adequate:	over 400 ng/l (300 pmol/l)
Marginal:	200-400 ng/l (150-300 pmol/l)
Deficient:	less than 200 ng/l (150 pmol/l)

Different assay systems produce somewhat different results and the laboratory may indicate slightly different values from these for deficiency.

(ii) Liver vitamin B12 concentrations.

Deficient:	less than 1 ug /g dry tissue

Serum and liver vitamin B12 concentrations are insensitive indicators of deficiency but good indices of excess. This has led to tests designed to measure the functional significance of vitamin B12.

(iii) Urinary metabolites - formiminoglutamic acid (FIGLU) and methyl malonic acid (MMA).

These accumulate in the urine in deficient lambs because vitamin B12 acts as a coenzyme in their break-down. FIGLU is not so valuable since pronounced clinical signs including actual loss of weight are necessary for it to accumulate whereas MMA starts to accumulate somewhat earlier. It may be possible to detect the elevation of MMA in the blood. (Urine can often be obtained in sheep by partial suffocation!)

In the food:

(i) Herbage - deficiency when below 0.08 mg Co/kg D.M.
(ii) Soil - deficiency when below 1.7 mg Co/kg D.M.

Treatment and prevention

There is no placental transfer of B12 and only low concentrations in milk; lambs require colostrum for their first supply.

1. Cobalt pellets (often called 'bullets')

   The most effective treatment is to give each lamb, at about 8 weeks of age, (not before, because they are too small to dose, and the rumen is not fully developed) a cobalt pellet (e.g. Permaco S, Coopers). This should be effective for at least a year but it costs approximately 50p per pellet, and for fattening lambs it may not be worth it, though it would be for ewe lambs which are to be kept for breeding.

2. Cobalt in soluble glass (see Cosecure, page 102)

3. Oral dosing with cobalt

   A drench can be made by dissolving 100g of cobalt sulphate in 2 litres of water to provide a stock solution. 250ml of this stock solution is then diluted in 5 litres of water; give young lambs 10ml (less than 0.5p), older lambs 15ml and adult ewes (where necessary) 30ml. This will need to be repeated every 3 weeks.

   C-Vet produce a preparation 'Copacobal' containing copper and cobalt which costs approximately 3p per dose and only 2 doses are advised (at 8 weeks and again in the autumn), but this seems to be an unusually long interval to be really effective.

An alternative is to dose with cobalt along with anthelmintic. Whilst cobalt and anthelmintic are often needed at the same time combined dosing does not always fit into a programme and care should be taken that one is not omitted because the other is unnecessary.

4. Cobalt supplementation in mineral mixture or water

It is unlikely that lambs will be on concentrates but if so, the mineral mixture should contain sufficient cobalt to raise the whole feed to 0.1mg Co/kg D.M. Cobalt supplementation of water supply is suitable only for sheep receiving piped water. Cobalt sulphate is dissolved and kept in a concentrated solution in a metering device plumbed into the water supply, in such a way as to deliver a known, steady amount of cobalt. The method is cheap, after initial installation of equipment, and animals need the minimum of handling.

5. Injection of vitamin B12

This will deal rapidly with an immediate problem but cost and the need for injections every 3 weeks precludes this as a preventive measure. There are several preparations and the dose for sheep is about 500 ug costing about 10p.

6. Application of cobalt sulphate to grazing land

Cobalt sulphate may be applied as a spray or as a granular top-dressing at 0.5 kg per hectare. Dressing need only be repeated every 3 years (or 6 years if the deficiency is only marginal) and only half the grazing need be treated, since the sheep graze the treated strip selectively. The cost of cobalt sulphate preparations varies markedly but will work out in the region of 15p per lamb.

PNEUMONIA

PASTEURELLOSIS

Significant advances have been made recently into the understanding of the epidemiology and control of pasteurellosis in sheep.

There are three different manifestations of the disease, all of which are associated with Pasteurella haemolytica:

(1) Pneumonia (Enzootic) - mainly in adults and particularly in the spring.

(2) Septicaemia and pneumonia - in lambs mainly in the spring and early summer.

(3) Septicaemia - in the fattening and store lamb in the autumn and winter, causing 'sudden death' (cf. Braxy).

Two types of the organism can be distinguished on their ability to ferment trehalose and Biotype A is mainly involved in 1 and 2 and Biotype T in 3.

There are at least 15 serotypes but 8 are responsible for the majority of clinical problems and these are included in the Moredun/Hoechst pasteurella vaccine Other infections such as PI3 virus, Jaagsiekte and T.B. fever may predispose to clinical outbreaks.

Clinical Signs

(1) The most obvious first sign is one (or a few) sheep found dead; Pasteurellosis is considered to be one of the most common causes of 'sudden death' in sheep.

(2) One (or a few others) may appear very ill and separated from the rest and unwilling to be driven; they are usually febrile (over 40°C, 104°F) and with laboured respirations. Auscultation may be convincing but often it is very difficult to be sure that pneumonia is the problem. Some may have crusted eyelids and nostrils and usually there is some coughing in the group. Up to 10% may become infected in an outbreak.

Diagnosis

(1) Try to find an excuse for the deaths/illness; in particular, enquire if the group has been moved or handled in the last day or so, or whether the weather has been noticeably different, e.g. wet and windy, or warm and still.

(2) Consider clostridial diseases, in particular Pulpy Kidney and Braxy, and check on vaccination dates.

(3) Necropsy of fresh carcases. <u>Pasteurella haemolytica</u> needs to be isolated in large numbers to be sure and ask the VIC for sero-typing to check against the vaccine sero-types.

Treatment and Control

(1) Rescue and isolate affected sheep. Treat these and as many contacts as seem worthwhile, with long-acting penicillin or oxytetracycline.

Sometimes the group is large and inaccessible, which means the incidence is unsure, PME is delayed and there is a very real difficulty in deciding when to do something, i.e. gather (which may make matters worse) and treat (the lot?). Often an 'irritating' number of deaths occur over a period of days or weeks and then unaccountably cease.

(2) Look at the environmental conditions e.g. wet, exposed and over-crowded, lush food; and consider moving the group (again?).

(3) Consider vaccination. The Moredun/Hoechst vaccine ('Ovipast') has a place both during outbreaks and in health programmes. This is a dead vaccine containing a number (but not all) of the sero-types, and requires two doses at approx. 4 weeks apart to provide useful protection (which clearly limits its value during an outbreak). If given to ewes in late pregnancy it not only affords protection to the ewes that spring (when the incidence of enzootic pneumonia is highest) but also to the lambs up to 3 to 4 weeks of age (but not 16 weeks as for clostridia) and so partially covers the first major risk period for pasteurella septicaemia. To cover the later period, lambs need to be given two doses of vaccine at 3 and 4 months.

However, pasteurellosis often occurs in lambs younger than this, and in such instances they should be vaccinated earlier; it seems that any colostrum derived antibody does not interfere with vaccinal response. The schedule matches the usual clostridial vaccine schedule, and to allow for this, the pasteurella vaccine has been combined with a 7 in 1 clostridial vaccine (Heptavac P) for breeding and a 4 in 1 clostridial vaccine (Ovivac P) for late fattening and store lambs.

Note that this vaccine should cover most, but not all, sero-types and that it is not cheap (approx. 20p per dose and up to 30p per dose if combined with clostridia) and note also that the outbreaks have the happy knack of dying out on their own!

PARASITIC PNEUMONIA ('Husk')

Although sheep are commonly infected with a number of nematodes which are found in various parts of the respiratory system the only species which is associated with clinical disease is Dictyocaulus filaria. It has a similar life cycle and epidemiology to D. viviparus in calves so that the heaviest pasture infections with infective larvae occur in September to November. These larvae can overwinter and act as a source of infection to the next season's lambs.

The disease is not so important clinically as in cattle which may be associated with the more frequent anthelmintic dosing of lambs for parasitic gastro-enteritis, which incidentally reduces the number of worms in the lungs. However, the worm is common and can cause coughing and loss in condition in lambs in August to October. The lambs develop a strong immunity and few worms are found in older sheep. Many lambs are sent to slaughter before the peak of infection.

Diagnosis

Diagnosis is based on clinical signs and seasonal incidence and can be confirmed by faecal examination for first stage larvae by the Baermann apparatus. Faeces should be fresh and D.filaria can be distinguished from the larvae of other non-pathogenic worms in that they are long (500um), have a blunt tail and a cephalic knob and contain refractile granules.

Treatment and Control

All the modern benzimidazoles, levamisole and ivomectin are effective against D.filaria and can be used to treat clinically affected lambs. Parasitic bronchitis is unlikely to occur where these drugs are used earlier in the year to control parasitic gastro-enteritis but some drugs (e.g. oxibendazole ('Widespec') and thiophanate ('Nemafax') are effective against intestinal worms but not against D.filaria at the usual dose.

CHRONIC ATYPICAL PNEUMONIA

This is an unsatisfactory name given to the type of pneumonia which closely resembles the pneumonia so commonly found in the fattening pig and housed calf. It is principally a disease problem in the housed fattening or store lamb (3 to 12 months old) and associated with Mycoplasma ovipneumoniae and perhaps other agents such as PI3 virus.

Clinical Signs

The disease is characterised by coughing and poor growth, plus the occasional acute pneumonic lamb (off food, listless, obvious respiratory distress and febrile) associated with secondary Pasteurella haemolytica infection. Infected lambs take several weeks longer to reach proper slaughter weight and consume more food to do so.

Diagnosis

Clinical signs and environmental conditions and the pathology of lungs at slaughter (similar to pig and calf).

Control

As for pigs and calves.

(1) More fresh air.

(2) Reduce numbers under one roof and in any one group.

(3) Split groups according to age and size.

(4) Inject with L/A tetracyclines or tylosin.

(5) Consider pasteurella vaccination.

SLOW VIRAL PNEUMONIAS

Jaagsiekte or Sheep Pulmonary Adenomatosis (SPA) = Driving Sickness

Although this condition has caused worrying losses in some flocks (particularly of Scottish origin) it usually only causes sporadic deaths, but its potential danger is illustrated by the losses that occurred in Iceland in housed sheep. It is a contagious adenomatous tumour of the lungs of sheep (only), thought to be associated with one or two slow viruses (herpes and retro) with an incubation period of up to 3 years.

Clincial Signs

An adult sheep (3 to 4 years) when driven, shows severe respiratory distress without much coughing, and over many weeks, the respiratory embarrassment ('bubbly porridge') increases and the sheep shows loss in weight despite antibiotic therapy. It is invariably fatal. Characteristically, if the ewe is held with its head lowered, relatively large volumes of frothy mucous exudate flow from the nostrils and this is infectious. There is no fever and the animal feeds normally up to the terminal stages (within a few weeks of first clinical signs). Jaagsiekte can predispose to pasteurellosis and so histology is required where pasteurellosis appears a persistent problem in adult sheep.

Diagnosis

There is no practical serological test (because the retro virus has not yet been isolated in cell culture) and diagnosis is based on clinical signs and PME. The lungs may weigh 4kg (9lb) (1.5kg (3lb) normal).

Control

Prompt slaughter of the obviously affected. At present, the condition should be regarded as a potential problem rather than a real one, but housing sheep could alter the situation. Once a serological test becomes available, then monitoring and pre-clinical culling becomes a possibility.

Maedi/Visna

For all practical purposes, this condition is similar to Jaagsiekte (although more recently introduced into the U.K. and only a few clinical cases have, as yet, been recorded). It is seen as a chronic progressive pneumonia affecting sheep (and goats) over 3 years old (Maedi = air hunger) and it is invariably fatal. Visna (wasting) is the nervous form of the disease, seen as a progressive paralysis of the hind legs of adults but it has not yet been recorded in this country. Both conditions are caused by the same retro-virus, and is spread particularly indoors and via the colostrum; a risk that needs monitoring because of the popularity of colostrum 'banks'.

The disease (like Jaagsiekte) has, as yet, little significance to the commercial lamb producer, but with the increased housing of ewes the position may change (like Jaagsiekte). However, (unlike Jaagsiekte) there are useful serological diagnostic tests which have led to the development of the MV - Accredited Flock's Scheme (for details contact the Ministry of Agriculture). This has particular application for the pedigree breeder, not only for control within his flock, but for registered sales and for shows.

Clinical Signs

Similar to Jaagsiekte but without the 'diagnostic' flow of chest fluid.

Diagnosis

Based on clinical signs and PME in the individual. It is most likely to be seen in an imported breed. Serological tests (Agar gel precipitin test (AGPT) and Enzyme - linked immunosorbent assay tests (ELISA) at £2.50 a test) are useful at the flock level, but false positive results (delay in production and immuno-suppression) do occur and make them less useful in the individual.

Control

Similar to Jaagsiekte.

(1) Slaughter the affected.

(2) Do not breed from the offspring of affected ewes.

(3) Keep a young flock and only keep replacements from the young ewes.

(4) Purchase from accredited flocks only.

(5) Join the Ministry scheme and eradicate by test and slaughter.

CEREAL PROBLEMS

UROLITHIASIS

This is mainly a hazard for the housed 2 to 4 month castrated male lamb fed a lot of concentrate. The fine sand-like calculi usually consisting of magnesium phosphate, obstruct the penis either at the vermiform appendage or at the sigmoid flexure.

Clinical Signs

(1) The lambs show discomfort with straining, kicking at the abdomen, twitching of the tail and general restlessness.

(2) A precipitate of crystals may be found on the preputial hairs which are often dry and sometimes bloodstained.

(3) 'Water belly' may occur, with urine leaking subcutaneously from a ruptured urethra, or into the abdominal cavity from a ruptured bladder.

Treatment (not easy, but urgent)

Depending on the severity and success -

(1) Spasmolytic and analgesic injection e.g. Buscopan composition (Boehringer), acetylpromazine (ACP C.Vet)

(2) Twice daily drenching (or tubing) with a litre of 0.9% ammonium or potassium chloride.

(3) Amputate the vermiform appendage with scissors (but in young lambs it is still adherent).

(4) Perineal urethrotomy.

Control

1. <u>When cases are occurring:-</u>

   (i) Change and reduce the amount of concentrate.

   (ii) Ensure plenty of fresh water.

   (iii) Supply salt (NaCl) - as licks, or <3% in ration or <10% in drinking water.

   (iv) Add 2% ammonium chloride to the ration.

2. <u>In future:-</u>

   (i) Ensure no magnesium added to concentrates (and not to exceed 200g MgO/tonne).

   (ii) Ensure Calcium/Phosphorus ratio is approx. 2:1; include 1.5% ground limestone in the diet.

   (iii) Ensure minimum of 1% salt in concentrate.

   (iv) Ensure plenty of fresh water.

   (v) Avoid castration.

RUMEN ACIDOSIS (Cereal overeating)

This frequently occurs when sheep are given too much cereal too quickly, particularly in the housed fattening or store lamb, but it occurs also in ewes when first housed and even when grazing barley stubble. Most frequently, there is just a short period of indigestion leading to scouring and dirty tails which self-cures. However, it can be much more serious with sheep just 'found dead' (yet another cause of 'sudden death' - try making a list!). Following the fermentation of excessive starch, the rumen contents become acid and hypertonic leading to rumenitis, scouring, dehydration and acidosis.

Clinical Signs

These appear about a day after excessive intake:-

1. A number look a bit miserable, with dirty tails and soft to watery faeces, and reduced appetite.

2. A few may be found dead or very ill, with diarrhoea, grinding teeth, sunken eyes, rapid respirations and recumbent.

Diagnosis

Is usually fairly obvious from seeing what has been fed and the clinical signs, but pulpy kidney and pasteurellosis should be considered and atypical chronic pneumonia may also be present in the group and this complicates the respiratory picture. PME shows an easily detached rumen epithelium with haemorrhagic sub-mucosa. The contents are sour-smelling and acid. (Use a narrow indicator paper - below 5 is suspicious and below 4.5 is certain.)

Treatment

Depending on how sick the lambs are:-

1. Drench with magnesium hydroxide mixture (10ml twice daily) or chalk (3g twice daily), and a bloat drench if tympanitic.

2. 1 or 2 litres of glucose saline i/v, 10ml of 2.5% solution of sodium bicarbonate, and 5ml 'Parentrovite'.

3. Oxtetracycline L.A. i/m (2-4ml).

4. Supply good quality hay and reduce the concentrates.

5. Supply fresh clean water at all times.

Future control

1. Introduce concentrates gradually over several days.

2. Provide adequate good quality hay.

3. Provide adequate trough space so that the greedy ones don't eat the lot.

4. Ensure adequate water supply.

EYES

CONTAGIOUS OPHTHALMIA

Contagious Conjuctivo - Keratitis (C.C.K.). Ovine Infectious Kerato-Conjunctivitis, (O.I.K.), Pink Eye, Heather Blindness.

This is a very common disturbance of sheep (and goats) eyes. Usually there is only superficial irritation recognisable by scleral congestion, lachrymation, blinking and some blepharo-spasm (grade 1). Some cases however,particularly older sheep show corneal inflammation with blood vessels and pannus spreading from the corneo-scleral margin (grade 2), and progressing to shallow corneal ulceration and temporary blindness (grade 3). Both eyes are usually affected but not always simultaneously or to the same grade.

The condition seems to be excited by handling and crowding, and is common in ewes in lambing pens. (Another disease likely to increase with housing). The disease is most probably caused by Mycoplasma conjunctivae but Chlamydia psittaci has been incriminated, and secondary bacteria make things worse, particularly perhaps staphylococci.

It is usually self-limiting but many carriers remain and relapses occur frequently. The duration of the disease is variable; if only the sclera is involved complete resolution may occur rapidly (in a day or so), but if the cornea is involved, inflammation may persist for 3 or 4 weeks. Blindness in adult ewes leads to starvation, pregnancy toxaemia and stray accidents.

Tetracyclines are the drugs of choice, using ophthalmic ointments. It is difficult to prescribe a proper schedule because most cases don't seem to warrant treatment, but grade 2 and 3 cases do deserve treating and this means that a farmer should have a tetracycline tube in his pocket when handling the flock, and the worst cases placed somewhere convenient and comfortable. Treating just the once is worthwhile although a course of twice daily for 3 days is indicated in the worst cases.

It is worth noting that this condition is very different in its aetiology, pathogenesis and severity, to so-called 'New Forest' disease in cattle, and that some ocular preparations, for example the penicillins for use in cattle, are inappropriate for sheep.

BRIGHT BLINDNESS

(Clear blind, Glass-eyed) (cf. progressive retinal atrophy (PRA) in dogs).

This is an irreversible blindness in adult sheep, mainly reported in Yorks/Lancs. flocks, but occurring in hill flocks in other areas. The condition affects both eyes equally and simultaneously. The retina shows progressive degeneration with atrophy of the rods and cones; the tapetal arteries and veins are narrowed and there is a marked green reflection from the tapetum. Ophthalmoscope examination is easy because the pupils are dilated and the cornea and lens are clear.

The prevalence may be over 5% and usually becomes apparent in the autumn following the bracken season. It is caused by a toxic factor in bracken; and may require several seasons of bracken consumption before signs become apparent - hence it is not seen in young sheep.

As the condition is slowly progressive, the sheep have time to adapt and the blindness may only become apparent when the sheep are driven or placed in strange surroundings. They are not unwell, appearing perhaps more alert than usual and are not easy to catch. They are, however, more prone to accident and to get lost. They require culling.

ENTROPION IN LAMBS

Entropion in new-born lambs is common, involving one or both lower eyelids. The incidence varies from flock to flock but cases are always being seen and many receive inadequate attention resulting in unnecessary discomfort and corneal damage. Some flocks have a distressing number of severe cases (<10%), others have none or only mild cases, and there is sometimes good evidence to suggest a breed factor, with a recessive gene involvement.

Treatment

1. Mild entropion can be easily everted by finger pressure, and that is all that's required.

2. More severe cases, where the lower-lid is turned-in a lot causing blepharospasm, keratitis and corneal ulceration, require more vigorous attention, such as:-

   (i) A subcutaneous injection of a liquid below the margin of the lid, sufficient to produce a bleb which everts the lid. Liquid paraffin is effective, but an antibiotic solution is more sensible.

   or (ii) Suture a tuck in the skin below the lid with 14 or 16 mm Michel clips (Arnold Vet. Products).

Most cases merit ophthalmic eye ointment at the time, and if possible for the next day or so.

Control

1. Consider the breeding lines.

2. Mark and record each case and don't breed from them.

SKIN DISEASES

The clinical signs of skin diseases in sheep may be:-

1. **Wool loss without pruritis or excessive scale formation**

   Check for causes of debility and/or stress at the time or in the month prior to the occurrence of wool break. Consider Border disease in young animals.

2. **Generalized or Regionalized Pruritis**

   (i) Scrapie

   Individual adult within flock - marked generalized pruritis (rubbing, nibbling and slapping lips). Skin may look normal with no excessive scaling, but often localized areas of skin damage and wool loss.

   (ii) Lice

   Irritation involving several or all members of flock, leading to wool loss and in severe infestations, debility. Signs most marked in winter months, no age incidence. Ectoparasites often difficult to find. Clip out a wadge of wool and examine in strong light and in laboratory. Irritation worse after flocking together. Dip all sheep and pick up wool debris in pens. One dipping may be sufficient. Spot treatment may possibly supercede dipping in the future.

   (iii) Keds

   Similar signs to louse infestation but irritation may be marked during summer as well as winter months.

   (iv) Sheep Scab

   Severe regional or generalized pruritis with an associated scaling, hyperaemia and wool loss leading to weight loss and debility and even death. Scaling may be marked sometimes, having a brown sugar appearance. Secondary Dermatophilus infection may be a complication.

Diagnosis rests on isolation and identification of the parasites from underside of crust (care must be taken to differentiate mange mites from forage mites). **N.B.** irritation increases when housed - get them in before you arrive in order to warm them up.

Sheep Scab is a notifiable disease under the Sheep Scab Order 1977

Any suspected outbreak must be reported immediately. If there are reasonable grounds for suspicion, the premises will be declared an infected place (Form A). All sheep must remain secured; no sheep may be moved on or off the premises without a licence. Previous movements of sheep from the infected place onto new premises will be traced. When sheep scab is suspected to be active in an area, the Minister may declare that area an Infected Area.

Dipping of Sheep

Mandatory dipping of all sheep will be required each year for the forseeable future, within dates dictated by the Ministry and modified in the light of experience gained in the previous year.

Authorised dipping must be carried out with a Ministry approved dip. The local authority must be informed giving 3 clear days notice and details of the proposed date, time, place, number of sheep and their source. Animals grazing Common Land must be subjected to co-ordinated dipping regime.

3. Marked Scale Formation, Minimal Irritation

Dermatophilus Infection

Thick scaly plaques involving either woolly areas of the body or confined to hairless areas. Debris and wool removable revealing either areas of shallow pink ulceration in active lesions or areas of healed thin epithelium in regressing lesions from which the wool or hair may be shed spontaneously.

The organism invades the skin when latter has high moisture content following thorough wetting, insect damage or skin injury.

Diagnosis rests on the clinical signs and on the identification of organism in Giemsa stained smears; must be differentiated from suent, canary wool, wool rot.

Clinical approach

1. Inspect the flock and note prevalence.

2. Note ectoparasites, shearing injuries and wetness and look for control of these.

3. Remove obviously affected wool and burn.

4. Consider streptomycin parenterally to worst affected.

5. Consider dipping in an anti-mycotic anti-parasitic wash, e.g. Young's Winter dip.

6. Consider powdering with potash alum.

7. Re-check in a fortnight.

4. Localised inflammatory lesions

Fly Strike

Remember that fly strike occurs suddenly and all good shepherds will examine their flocks at least once daily during the season so that all new cases may be identified quickly. Often the sheep are separated from the rest of the flock.

(i) Remove matted fleece, clean up, apply BHC cream and if necessary parenteral antibiotics (Ivomectin may be of value). Some need slaughtering.

(ii) Remove triggers e.g. worms - leading to scouring, crutching of animals grazing lush pasture may be indicated.

(iii) Dip all the sheep. Many dips have a colour marker which, when it fades, indicates that dipping needs repeating. Frequency will be dictated by rainfall, type of dip and breed of sheep.

Head Fly (Hydrotaea irritans)

A non-biting fly that swarms round the heads of sheep and other species causing great annoyance, interrupting grazing and leading to self-inflicted traumatic lesions which may become infected or subject to fly strike and causing economic loss. Horned sheep are usually but not invariably involved. The fly breeds in litter and damp vegetation in the early part of the year, the adult flies emerging in June and July and remaining active until September or October. Control is difficult. Distancing lowland flocks at least 15 metres from wooded areas and large hedges reduces the problem. Changing breeds avoiding horned and black-faced breeds helps but is usually impractical. Head caps have been tried. Organophosphorus (O.P.) creams applied at least every 14 days to the face and ears helps but is expensive and time consuming. O.P. or pyrethroid impregnated ear tags may offer a solution in the future.

Pyrethenoid based pour on preparations have recently produced encouraging results and may yield an effective method of control in the future.

Periorbital dermatitis

From bruising at troughs, often in housed sheep. Separate, clean up, antibiotic spray and injection and provide more trough space (60cm (2')/sheep). Although staphylococci are often serious primary or secondary invaders other organisms such as Pseudomonas can cause problems.

Posthitis (Pizzle rot and ulcerative dermatosis)

A condition causing ulceration and necrosis of the sheath and sometimes tip of the glans penis and vulvo-vaginitis in ewes. A viral cause has been suspected and *Corynebacterium renale* infection is often present, together with *Pseudomonas* as a complication. It is most frequently noted in flocks fed on rich leguminous pastures. Infected rams should be isolated and lesions treated with oxytetracycline creams. Ewes showing active lesions should not be tupped - infected animals should be culled when possible.

Ringworm

Uncommon, but is sometimes seen in housed sheep particularly if shorn. Usually *Trichophyton verrucosum*, and often associated with the use of buildings previously used to house cattle. Treat as for cattle. Separate obviously affected, remove scabs and spray tincture of iodine (navel spray). Usually improve rapidly when turned out. Griseofulven is of doubtful value and must not be fed to pregnant sheep.

Photosensitization

Initially produces a non-painful oedema of the head and in particular the ears (the latter often becoming pendulous) and over non-woolly parts of the legs. Animals are not critically ill unless the condition is associated with liver damage. In the latter stages skin necrosis may occur, e.g. tips of ears may be lost. Should the condition appear immediately after shearing, lesions will be widespread over all exposed areas of skin. Place in a dark box or shed; prevent further contact with the sensitizing agent. Give steroids and when necessary antibiotic cover. Spray lesions with terramycin spray containing gentian violet. Prognosis good provided further contact with the exciting cause can be prevented and there has been no liver involvement.

Orf (Contagious Pustular Dermatitis, Scabby Mouth)

This is an important condition of sheep and goats. Caused by a pox virus (easily identified by E.M.) which may persist in skin debris and lambing areas for a long time but carrier sheep are probably more important. It is essentially a contagious disease and requires close contact and possibly epithelial drainage to be present.

It causes a granulomatous condition of the lips, sometimes spreading over other areas of the head and limbs in lambs and infecting the skin of the udder and teats in ewes. This may result in lambs having difficulty in sucking, leading to loss of condition and sometimes to mortality but the most important clinical effect is mastitis in ewes.

The virus has also been found in conjunction with Dermatophilus in cases of strawberry foot rot, a condition involving the lower part of the limbs typified by sores and swelling over affected areas, and lameness.

Treatment of orf consists of treating the lesions locally with terramycin spray to control secondary infection, giving supplementary feeding and relying on natural resolutions of lesions, which normally takes place within three to four weeks.

Treatment of strawberry foot rot is often unsatisfactory and culling of infected animals may be required.

In serious outbreaks, transfer all non-affected animals to clean premises and ban the use of those lambing pens for next year. FAM disinfectants may be of value in the control of the viral contamination of buildings.

Two vaccines are available, one having a longer shelf life but they should only be employed when there is evidence of the disease on the farm and when other methods have failed, as the control they offer is not absolute. Immunity conferred is incomplete and is believed to be a cutaneous cell mediated response and although rained concentrations of IgG may be detected in blood and lymphoid tissue, this is non-protective and passive transfer of immunity in colostrum does not occur. There may also be strains with different antigens. Although vaccination may shorten recovery time it is of doubtful value during an outbreak; however, one accepted placed for vaccination is annually in early or mid-pregnancy in an attempt to avoid orf on teats, the lesion which causes the greatest problem.

Care must be taken in handling animals from infected flocks as the virus can infect man producing hard, painful, circumscribed lesions on the hands or face which persist for four to six weeks. It may also cause lesions on cows teats similar if not identical to Pseudo-cowpox (Paravaccinia).

SHEEP HEALTH SCHEME

Many sheep farmers are now requesting advice from their veterinary surgeons in the formulation of health programmes. We reproduce here our approach to this opportunity, which is similar to those produced by others (see Sheep Veterinary Society).

OBJECTIVE: To produce, monitor and maintain a health scheme, in the belief that this will both decrease disease and increase production.

THE SCHEME: The scheme has three components:-

1. A written Health Programme containing recommendations for the control of expected diseases and to improve production.

2. Planned visits (3-6 annually) to monitor the health and production.

3. Reactions to events and new advances during the year, to keep the programme up to date.

Its success hinges on the degree of co-operation between the farmer and the practising veterinary surgeon and other advisers, and requires honest pooling of information.

1. Health Programme

This programme is in the form of a diary and is tailor-made to suit the needs of the farm following full discussion at the first visit. It is based around the farmer's proposed dates for tupping, housing, lambing and weaning, and includes dates for some or all of the following:-

(a) Tupping - including early breeding, synchronisation, AI and ram testing.

(b) Condition scoring of ewes.

(c) Vaccination against clostridia, orf, abortion, foot rot, pasteurellosis.

(d) Dosing for copper, cobalt and selenium/Vit. E. deficiencies.

(e) Worming for roundworms.

(f) Fluke drenching.

(g) Treatment for coccidiosis.

(h) Tapeworm treatment of dogs.

(i) Feeding and food analysis.

2. **Planned Visits**

Several visits will be needed in order to assess the application and usefulness of the Health Programme. In the first year, six visits will usually be necessary according to the following plan; in subsequent years fewer visits and samples are likely to be required.

	DATE OF VISIT (approx.)	JOBS (some or all)	TIME (approx.)
VISIT 1*	Two months before tupping	1. Rams (all): Condition score, check fertility and feet 2. Ewes (100 selected): Condition score, cull (teeth, age, udders) 3. Store lambs (a selection) Condition score 4. Take samples of blood and faeces for copper, B12, selenium, fluke and worms and abortion profile (Toxoplasma and Enzootic) 5. Check hay or silage and take samples for fibre, protein and energy estimates 6. Inspect pasture, examine for snail habitats 7. Complete Farm Questionnaire (sent in advance) 8. Discuss 'clean grazing' 9. Prepare draft programme following this visit and receipt of lab. results	4 hours
VISIT 2	Approximately 1 month later	Discuss draft programme with farmer and finalise programme	2 hours
VISIT 3*	6 weeks before lambing	Condition Score 100 ewes. Assist copper injections and fluke drenching Advise about feeding (including trace elements) Treat clinical cases.	2 hours
VISIT 4	2 weeks before lambing	Condition score 100 ewes. Assist Clostridial vaccinations and worming Advise about housing facilities, colostrum, lambing and recording Treat clinical cases	2 hours
VISIT 5*	At lambing time	Clinical events Recording of lambs Advise about hypothermia and starvation *E.coli*, coccidia and worming	2 hours
VISIT 6	12 weeks after lambing	Assist clostridial and pasteurella vaccination, and worming of lambs Recording	2 hours
			TOTAL: 14 hours

* Visits in subsequent years

N.B.

When more than one flock is involved with widely different lambing dates, more than 6 visits will be necessary, although it is probable that some can be combined. In addition to the 6 planned visits, advice will be available by phone.

It is useful to have a check list, particularly for the first visit, along the following lines

SHEEP FARM VISIT - CHECK LIST

FARM/FARMER	OBSERVATIONS
1. Geography	
2. Geology	
3. Climate	
4. Possibilities for - Ticks	
Fluke	
5. Economic position:	
Income from Sheep	
Income from other sources	
Gross margin/ewe/ha	
Gross margin/ha	
6. Drugs and instruments (have a good look round)	
7. Disease control measures (what's wrong with them)	
8. Welfare/care of sick	
9. Labour/dogs (particularly at lambing time)	
10. Use of Vets	
ADAS	
Commercial Reps	
ATB	

THE SHEEP	OBSERVATIONS

1. General look at each flock/ age group

Number of flocks

Variety of breeds

Extensive/Intensive

General condition

Lameness

Coughing

Dirty tails

Wool break

2. Examine (handle)

1. 10% of each flock/age group

Condition Score

Teeth/Age

Blood sample (if necessary)

(a) Abortion 'profile'

(b) Deficiencies

Faecal sample (if necessary)

(a) Fluke

(b) Worms

(c) Coccidia

2. Clinical cases (systematic & sample)

e.g. Very thin

Very lame

Very sick

Very nervous/depressed

Very 'moth-eaten'

Very dirty-tailed

Tail end lambs

THE SHEEP	OBSERVATIONS
3. <u>The Rams</u> (All)	
Condition Score	
Feet	
Genitalia	
Faeces (sample)	
4. <u>Abortions</u>	
Samples	
5. <u>Suitable carcases</u>	
For PME.	

	THE FOOD	OBSERVATIONS
1.	Stocking rate (ewes/ha)	
2.	Quantity and quality of grazing	
3.	Reclamation/fertilisation	
4.	History of deficiency diseases	
5.	Mineral supplementation (food and dosing)	
6.	Stored roughage (hay, silage) Quantity/quality	
7.	Concentrates Menu and quantity per ewe/day	
8.	Storage (& cats!)	
9.	Rack and trough space/ewe	
10.	Colostrum (source & quantity)	

	THE SHELTER	OBSERVATIONS
1.	State of repair	
2.	Size/space/stocking density (ewes/cu.m)	
3.	Atmosphere	
4.	Bedding	
5.	Water supply	
6.	Lighting	
7.	Trough space (ewes/cm.(inches))	
8.	Lambing pens -	
	numbers	
	cleanliness ("Squelch" etc)	
	heating	
	risks	
	Lambing and revival kit	
	Drugs	
9.	The Sheep -	
	puffing/blowing/coughing	
	'sweating' (if not shorn)	
	Shivering (if shorn)	
	cleanliness	
	lameness	
10.	Isolation facilities	
	e.g. abortions/sick	

3. Reactions to events

It may be necessary to modify the original Health Programme from time to time during the year, following information obtained during and in between the planned visits. Information obtained about some or all of the following will be carefully considered:-

1. Breeds, ages and flock numbers.
2. Condition of sheep.
3. Quality and quantity of food.
4. Housing arrangements.
5. Lambing arrangements.
6. Survey of snail habitats and drainage schemes.
7. Pasture and grazing management.
8. Clinical disease e.g. Pregnancy toxaemia, abortions, lameness, lamb losses, pneumonia.
9. Laboratory findings, including post-mortem examination of a sufficient number of carcases to establish the cause of disease.

The programme will need amendment at the end of each year to incorporate new research, drugs, vaccines and in the light of the experience of the past year and the alteration in the aims of the farmer.

POSSIBLE CHARGES

First Year (October 1986 Prices)

1.	For 6 visits	14 hours @ £25 per hour		£350.00
2.	Laboratory (as required)	12 bloods for mineral trace element profile	£ 70.00	
		12 faeces for fluke	£ 36.00	
		12 faeces for roundworm eggs, coccidia	£ 42.00	
		12 separate necropsies	£120.00	
		2 food analyses	£ 40.00	

FARM QUESTIONNAIRE (Sent in advance of Visit 1 and thus completed in June - August, depending on tupping dates)

FARMER'S NAME:

ADDRESS:

TELEPHONE:

VETERINARY SURGEON:

NAME/ADDRESS/TELEPHONE NO.

GENERAL INFORMATION

Total Area (ha.)	
Total grassland	
Silage	
Hay	
Roots - Turnips	
Home produced Oats/Barley for sheep	

Other stock beside sheep	
Labour available	
Tenant/Owner	

Is there a large scale map of the farm available	

FLOCK INFORMATION

Total No. of Ewes [ ] No. of Flocks [ ]

(A flock is a group of sheep with an identity! i.e. lambing date, pedigree, special purpose. If a group of sheep has a planned lambing date a month or more apart, treat as a separate flock.) Since a sheep year normally commences at tupping (Summer/Autumn) and concludes with lamb sales about 12 months later, the information requested should refer to the ewes and their lamb crop, i.e. ewe numbers in <u>last</u> calendar year and lamb numbers <u>this</u> calendar year.

	Flock 1	Flock 2	Flock 3
Breed of ewe			
No. of ewes excluding ewe-lambs			
Ewe lambs: No. / Age at tupping			
Breed of Rams			
No. of Rams			
Tupping date			
Lambing date			
Shearing dates			
Dipping dates			
Weaning dates			
Housing dates			
Wintered away dates			
Ewe feeding: What/When/How much			
Dock/Castrate: How/When			
Where lambed			
Is clean grazing available at turn-out i.e. not grazed by sheep the year before			
Is clean grazing available in July i.e. not grazed by sheep this year			
Lamb sales, When/Where			
Purpose of flock			
Special practices, (e.g. synchronisation, A.I., Scanning)			
Ewes purchased: Age/Source			
Percentage of flock lambed over 4 week period			

VETERINARY JOBS

DISEASE CONTROL

VACCINATION			
	With What	To What	When
Clostridia (& pasteurella) Foot Rot Orf Any Others			
MINERALS			
Copper Cobalt Selenium			
PARASITES			
Worms Fluke Coccidia			
OTHERS			
Footbathing Condition Scoring Tape-worm dosing to dogs			
CLINICAL PROBLEMS ENCOUNTERED			
Ewes Lambs Laboratory reports			
What do you think is your most important problem?			

PRODUCTION FIGURES

LAMBS	FLOCK 1	FLOCK 2	FLOCK 3
Number of lambs			
Born alive			
Born dead			
Dying in first 7 days			
Sold for slaughter before weaning			
Weaned			
Sold for slaughter after weaning			
Sold as stores			
Retained as stores			

Lambs sold for slaughter			
Month	Number	Av. e.d.c.w. in kg.	Av. price (state whether including variable premium)

EWES	Number	When	Why
Died			
Culled			
Sold			
Purchased			

AN EXAMPLE OF A HEALTH PROGRAMME (an expensive one!)

	FLOCK (1) 85 ewes	FLOCK (2) 33 ewe lambs
VISIT 1 September	Check ewes and rams. condition score & feed, feet, footbath, Footvax & culling.	
10th October	TUP	Check ewes and rams. Condition score & feed, feet, footbath, Footvax & Cosecure. Culling. 5th TUP
November		
December	Ewes - concentrate with monensin, Footvax and footbath	
VISIT 2 January	Ewes - house, shear, bath, worm and condition score	Ewes - concentrate with monensin, Footvax and footbath
VISIT 3 February	Ewes and Rams - Heptavac P Condition score. Lambing kit and pens	Ewes - house, shear, foot-bath & worm & condition score
VISIT 4 March	LAMBING Lambs - monensin in creep from 2 weeks	Ewes - Heptavac P. Condition score. Lambing kit & pens
VISIT 5 April	TURN OUT Ewes - Footvax 3 weeks	LAMBING TURN OUT Lambs - monensin in creep from 2 weeks Ewes - Footvax
May	Ewes and lambs - worm 3 weeks	3 weeks Ewes and lambs - worm
VISIT 6 June	Ewes and lambs - worm 3 weeks Lambs - worm, Heptavac P and Footvax	3 weeks Ewes and lambs - worm
July	WEAN	Lambs - Heptavac P & Footvax
August	Lambs - Heptavac P & Footvax	WEAN Lambs - Heptavac P & Footvax

SYSTEMATIC EXAMINATION OF A SICK EWE - A CHECK LIST

(In order, and stop when necessary)

1. FROM DISTANCE BEFORE HANDLING

Grade of Illness:	1. Mild, Severe, Dying 2. Acute, Chronic
General appearance:	Normal, Excitable, Depressed
Fleece:	Normal, 'Bald' patches, Dirty tail, Rubbing
Condition:	Fat, Thin or Average
Faeces	Normal, Loose, Absent
Respirations:	Rate, Depth, Cough, Discharges
Head:	Tilt/Aversion, Facial paralysis, Dribbling, Lachrymation, Blepharospasm (one or both eyes)
Feeding:	Normal, Hungry, Off food, With difficulty
Mobility:	Ataxic, Recumbent, Lame, Circles

---

2. ON HANDLING

Difficult or easy to catch

Condition Score:	1 - 5
Fleece/Skin:	Wool break, Ectoparasites, Lumpy, Itchy (rub test)
Age/Teeth	Incisors, Molars, Ramus

Eyes:	Blind (one or both, Menace test), Cornea(s) clear or cloudy, Sclera pale or congested
Cranial Nerves:	V & VII - Movement and sensation (eyelids, face, lips, gums)
Temperature:	Above 40°C (104°F)
Pulse: (Femoral)	Strong/Weak, V. rapid/Recovery
Faeces:	Pellets, Soft, Scour, Smell
Vulva:	Discharge, Membranes, Smell
Pregnancy:	Abdominal distension, Ballot
Udder:	Swelling (2A, 2C or 3), Milk, Teats
Feet:	Interdigital, Overgrowth, Under-running and separation, Smell, Pain (one or both claws), Coronary band swelling
Joints/Limbs:	Carpus, Hock, Stifle, Hip Pain, Swelling, Stiff, Abrasions, Wasting
Auscultation:	Heart - rate and sounds, Lungs - bubbles and squeaks, Rumen movements
Urine:	Colour, Ketones

SHEEP VETERINARY SOCIETY

This is a specialist division of the British Veterinary Association and membership is open to all members of the veterinary profession (whether BVA members or not) and others interested in sheep. The Society holds two meetings a year (normally 2 or 3 days) in April and September which include farm visits as well as lectures and discussions in a friendly atmosphere. Membership is a 'must' for all who wish to keep up with developments in sheep veterinary work.

The Secretary, Mr. Chris Lewis, Fields Farm, Green Lane, Audlem, Cheshire,is always pleased to hear from potential members. The annual membership fee is £5.00.

FURTHER READING (for the lost sheep!)

1. Blood, D.C., Radostits, O.M. & Henderson, J.A. 1983. Veterinary Medicine, Sixth Edition. Bailliere Tindall, London.

2. Diseases of Sheep, Ed. W.B. Martin, 1983. Blackwell Scientific Publications.

3. Speedy, A.W. 1980. Sheep Production: Science into practice. Longman Handbooks in Agriculture.

4. Haresign, W., McLeod, B.J., Webster, G.M. 1983. Endocrine control of reproduction in the ewe. In: Sheep Production, Ed. W. Haresign. Butterworth, London.

INDEX

-A-

-B-

-C-

-D-

-E-

-F-

-G-

-H-

-I-

-J-

-K-

-L-

-M-

-N-

-O-

-P-

-Q-

-R-

-S-

-T-